56TH FIGHTER GROUP

By Larry Davis
Illustrated by Don Greer
and Perry Manley

 **squadron/signal publications**

From the P-47 Thunderbolt of the Second World War to the F-16C Fighting Falcon of today, the 56th has been in the forefront of U.S. Aviation technology.

ISBN 0-89747-240-3

If you have any photographs of the aircraft, armor, soldiers or ships of any nation, particularly wartime snapshots, why not share them with us and help make Squadron/Signal's books all the more interesting and complete in the future. Any photograph sent to us will be copied and the original returned. The donor will be fully credited for any photos used. Please send them to:

Squadron/Signal Publications, Inc.
1115 Crowley Drive.
Carrollton, TX 75011-5010.

Acknowledgements

Air Force Museum
CMSGT Tom Brewer, USAF Ret.
Robert Cavanaugh
CAPT Norm Crocker, USAF Ret.
Jeffery Ethell
TSGT Thomas Fitton, USAF
Bart Ingram
BGEN Mike Jackson, USAF Ret.
Fred LePage
MSGT David Menard, USAF Ret.
LTC Barry Miller, USAF
Wayne Mutza
Mick Roth
Dr. Abraham Scherr, PHD.
Paul Vercammen
COL Hubert Zemke, USAF Ret.

TSGT Frank Beuneman, USAF Ret.
Ken Buchanen
COL Harold Comstock, USAF Ret.
Robert F. Dorr
56th TTW/History Office
Don Garrett, Jr.
Marty Isham
Art Kreiger
Ernie McDowell
Joe Michaels
CAPT J.J. Miller, USAF Ret.
MSGT Merle Olmsted, USAF Ret.
"Stormy" Sadowski, USAAF Ret.
CAPT John Truluck, USAAF Ret.
SGT Ed Wilds, USAAF Ret.

(Overleaf)
A P-47C Thunderbolt of the 56th Fighter Group over Long Island during the Summer of 1942. The 56th FG combined P-47 aircraft test flights with combat training. The mission was the defense of the Republic Aviation plant. (USAF)

Introduction

How do you spell "air superiority?" During the Second World War the USAAF spelled it: Z-E-M-K-E. The 56th Fighter Group, commonly known as "Zemke's Wolfpack" (after their commander, COL Hubert Zemke) had cut a wide swath through the Luftwaffe from the very beginning. By the time the war ended the 56th was the top scoring Fighter Group in the 8th Air Force with a total of 665 1/2 air-to-air victories. There were forty-two air-to-air aces in the 56th, plus seven air-to-ground aces, for a total of forty-nine men with the title "Ace," more than any other unit in the entire U.S. Army Air Force. Two of these were the top scoring aces in the European Theater. The Luftwaffe had learned to believe the motto of the 56th - *Cave Tonitrum* —Beware The Thunderbolt!

When the war ended, many Army Air Force units with distinguished wartime records were disbanded during the post-war force reduction. The 56th was not one of these. In fact, they were one of the first units to transition to jet fighters. They were the first jet fighter unit to cross the Atlantic as part of a show of force for the Russians during the Berlin Blockade. Although the 56th did not participate, as a Group, in the Korean War, they sent many pilots to units that were fighting there. The mission of the 56th during that war was the air defense of the United States. This remained their mission until 1967 when they again went to war in a little Asian country called Vietnam.

During the Vietnam War, the 56th left their jets behind and once again flew propeller aircraft in some of the strangest missions in that strange conflict. The 56th was a Special Operations unit, flying B-26s, C-47s, T-28s and A-1s. Their missions ranged from truck-busting on the Ho Chi Minh Trail, to air rescue deep inside North Vietnam. The 56th SOW added thirteen more campaign streamers to their flag during the Vietnam War.

Following their second war, the 56th made the transition back to jets, this time as a Tactical Fighter Wing flying F-4 Phantoms. Today the 56th is responsible for training air and ground crews for Tactical Air Command units in the F-16 Fighting Falcon. This, then, is the story of the 56th Fighter Group, from the Second World War until today. It is the story of dedicated people — dedicated to freedom and willing to fight for it wherever and whenever it becomes necessary. The men and women of the 56th are and always have been — Ready and Waiting.

When first formed, the 56th Pursuit Group had a mixed inventory of P-35, P-36, P-39 and P-40 aircraft and were slated to begin training on the Lockheed P-38. CAPTs Dave Schilling, G. Gorrell, Sylvester Burke, Lucian Dade, Phil Tukey and S. Blair pose with the mascot of the 63rd PS, "Snafu," at Charlotte Airport during 1941. (Lucian Dade)

The Beginning

The year was 1940. Across the Atlantic, Hitler's armies had conquered most of Europe with England, Russia and the Suez Canal next on their list. Japan's Greater East Asia Co-Prosperity Sphere was about to put into motion the events that would eventually lead to the "Day Of Infamy." In the United States, the Congress authorized, as part of the National Selective Service Act, the expansion of the U.S. Army Air Corps. The USAAC was expanded with a number of additional air groups and one of these new groups was the 56th Pursuit Group.

The 56th PG was formed on 20 November 1940, but it was not officially activated, at Savannah Air Base, Georgia, until 14 January 1941. On that morning three officers and 150 enlisted men reported for duty. They had been drawn from the ranks of the 27th Bomb Group and were used to form the three squadrons within the Group: the 61st, 62nd, and 63rd Pursuit Squadrons.

At this time, the Army Air Corps did not have enough aircraft and equipment for the new squadrons within the expanding Air Corps and by the Summer of 1941, the Group, now based at Charlotte Army Air Base, had an inventory of three P-39C Airacobras, five P-40B Tomahawks and a few trainers. The Carolina Maneuvers, however, would suddenly change all of this.

The Carolina Maneuvers were the biggest peace time exercise in Army history. Some 175 "enemy" bombers were to attack targets

Betty was a P-47C assigned to the 61st Fighter Squadron during October of 1942. The Thunderbolt had a distinctive "milk bottle" shape which led crews to nickname the aircraft the "Jug." The band around the rear fuselage is Red indicating that this P-47 is flown by a Flight Commander. (AFM)

throughout the Carolinas. These targets would be defended by several of the newly formed pursuit groups, including the 56th PG. For the maneuvers, the Army delivered ten new production P-39Cs to the 56th. The "raid" took place on 24 October 1941, with the enemy forces being repulsed. The extensive training the unit had received at the Myrtle Beach Gunnery Range had proven its worth. The 56th at least appeared to be a viable fighting force, although the Group could hardly be called combat ready.

With the Japanese attack on Pearl Harbor, the war "games" ceased; now it was for real! On 10 December 1941, the Group Headquarters and the 61st FS moved to Charleston Airport, the 62nd FS moved to Wilmington and the 63rd FS went to Myrtle Beach. Their mission was air defense of the Southeastern United States. Additional P-39Cs and some Curtiss P-36 Hawks were added to the group to bring them up to strength.

The Group became part of the New York Air Defense Wing a year after being activated. On 15 January 1942, the 61st FS moved to Bridgeport Municipal Airport, the 62nd to Bendix Airfield in New Jersey, and the 63rd was given the responsibility for the defense of what would eventually become an important facility in Group history: the Farmingdale, Long Island, Airfield. This was the site of the Republic Aviation factory. During June of 1942, the Group began their long association with the Republic P-47 Thunderbolt that would last throughout the Second World War.

The 56th had two Group Commanders, COL David Graves and COL John Crosswaithe, before finally meeting the man that would lead them into combat. On 16 September 1942, MAJ Hubert "Hub" Zemke took over as Group Commander. He had recently returned from a special liaison mission to England and the Soviet Union, where he observed the air war against the Luftwaffe first hand. His immediate task was to prepare the Group for combat operations against the, as yet undefeated, Luftwaffe.

Zemke began a rigorous training schedule with their new aircraft, which still had a number of the usual bugs. This training took its toll as eighteen pilots lost their lives before ever facing the enemy. But the rigorous training came just in time since the Group was reasonably ready when they were alerted for overseas movement on Thanksgiving Day 1942. At 0300 on 6 January 1943, the Group, some 800 men strong, boarded the HMS QUEEN ELIZABETH, disembarking on 12 January 1943 in Great Britain.

Their next stop was an RAF airfield at Kingscliffe where the Group received a number of brand new Republic P-47C Thunderbolts on 22 January. Training began immediately with and against veteran Royal Air Force pilots. The Group learned a lot from the British pilots that had beaten the Luftwaffe some three years earlier. On 5 April 1943 the Group moved to Horsham Saint Faith for the ultimate test - COMBAT!

The first Thunderbolts arrived in England during January of 1943 and were assembled and tested at RAF Goxhill. This P-47 is being readied for delivery to the first P-47 combat unit, the 4th Fighter Group at Debden. Although the 56th FG was the sole Army Air Force unit trained in the P-47, they were the last to receive their aircraft, since the 4th and 78th Groups had priority. (USAF)

Combat

The move to Horsham Saint Faith meant that the 56th was now considered combat ready. In fact, four 56th pilots, LTC Zemke, MAJ David Schilling, MAJ Loren McCollem and MAJ Phil Tukey had flown their first combat mission on 8 April 1943 while attached to the 4th FG. The mission was a fighter sweep over the Pas-de-Calais. On 13 April the 56th FG flew its first combat mission over France.

The mission involved all the units of VIII Fighter Command, one squadron each from the 4th, 56th, and 78th Fighter Groups. COL Zemke led the 56th, with four pilots from each squadron taking part. The mission was a fighter sweep, or "Rodeo," to Saint Omer. The objective was to entice the Yellow-nosed Abbeville Me 109s into a fight. Zemke's P-47 developed an oxygen problem and he was forced to turn back, leaving MAJ Schilling in the lead. But the Abbeville Boys would not come up and the Group returned home. The mission, however, did cost the Group its first aircraft loss. CAPT Dyar's engine froze over France and he had to glide back across the Channel, crashing in a field just five miles from the sea.

The Luftwaffe finally came up on 29 April during a Rodeo to Woensdrecht, Holland. MAJ Schilling led all three squadrons in their first actual combat. Near Antwerp, the Group was attacked by about twenty Me 109s and Fw 190s. Radio problems with Schilling's aircraft kept the top two squadrons from knowing about the fight going on below them. One hour and forty-eight minutes after takeoff, the Group returned to base, minus two aircraft and their pilots. LT Winston Garth had engine trouble and was forced to bail out over the Channel. CAPT John McClure bailed out over Holland. Both were captured by the Germans.

The first "Ramrod," or bomber escort mission, came on 4 May when the 4th and 56th FGs escorted fifty-four B-17s to the Ford plant in Antwerp. Again there were no results, but one RAF Spitfire pilot was very grateful that the Group was around. LT Walker Mahurin took his flight down to engage four Fw 190s that were closing on a lone Spitfire. Mahurin overshot the German fighters, but caused the Germans to break off their attack. The Group flew its first "Circus" on 19 May. A "Circus" utilized a small bomber force with a large fighter escort to lure the Germans into a fight. The mission was another Milk Run. One flight of Fw 190s jumped the 61st Squadron but ran for home before anyone could get a shot. May of 1943 ended with no kills for the Group.

On 12 June, CAPT Walter Cook, leading Yellow Flight of the 62nd Squadron on a Rodeo to Roulers, France, shot down an Fw-190. Near Ypres, France, the Group had spotted a gaggle of Fw 190s about 5,000 feet below them. When Blue Flight attacked, the Germans split, four making a 180 degree turn to get behind the Thunderbolts. Cook watched the action until the Fw 190s were almost in position, then took his flight down to attack. Closing to within 300 yards, his first burst hit the left wing and fuselage of the Fw 190, exploding one of the 20MM ammunition bays. The wing fell apart and the Fw 190

A P-47C of the 62nd FS parked on the grass at Horsham Saint Faith Airfield during June of 1943. At this time the aircraft carried no unit codes, only a plane-in-group number and White recognition bands on the nose and tail surfaces. (USAF)

CAPT Horace "Pappy" Craig (LM-R) leads a formation of five Thunderbolts of the 62nd FS during the Fall of 1943. The White nose and tail bands were added to all Thunderbolts due to a combat recognition problem between the P-47 and the German Fw 190 fighter. (Fred LePage)

went into an uncontrolled spin, disappearing into the cloud deck below. The Group had its first confirmed kill.

The following day, on a Rodeo to Ypres, COL Zemke shot down two Germans and LT Robert S. Johnson got his first. By the end of June the Group had six kills, but they had also lost five of their own.

On 26 June the Group lost four Thunderbolts and their pilots during a Ramrod to Villacoubly. Bob Johnson was Shaker Yellow Four. He and the rest of the Group were at 27,000 feet when they spotted a large formation of Fw 190s about 1,000 feet below them. Johnson started to call a warning but the Germans jumped all over the Group before anyone could react. Johnson both heard and felt the cannon shells hitting his P-47, which was thrown down and to the right. The engine started to smoke and bang like someone beating on a garbage can and smoke began to fill the cockpit.

Johnson decided to leave the crippled aircraft. He started to slide the canopy open, but it halted just nine inches back. No matter how hard he tried, Johnson could not open the canopy far enough to get out. He even tried putting his feet on the instrument panel and shoving with all his strength. Johnson pulled the emergency release handle, which should have jettisoned the canopy. But all that happened was that the canopy glass blew out, although most of it had already been blown out by German cannon shells.

Johnson was down to 19,000 feet when the engine fire went out and the smoke cleared from the cockpit. He still could not see because of the oil on the windscreen. Turning north, Johnson began the long flight back to England. The engine would only run at idle, but the bird was still flying. Johnson hoped he could make the Channel and maybe ditch next to a British ship.

At 8,000 feet, Johnson was approaching the Channel when a lone Fw 190 pulled in close to have a look at the crippled Thunderbolt. The German pilot pulled to within fifty feet and stayed there. Fearing that the German would get in really tight before opening fire, Johnson turned into him as if attempting to get into a firing position. The German immediately pulled the Yellow-nosed Fw 190 up, dropped back a few more feet and pulled in behind Johnson.

Johnson felt the P-47 shudder as the German opened fire at point blank range. Johnson could do nothing except crouch behind his armored seat back and use the rudder to throw off the German's aim. As the Fw pilot flashed by, Johnson would fire a few rounds. Finally

the bewildered German pulled up directly alongside Johnson, flying formation with the Thunderbolt as if trying to figure out what was keeping the P-47 in the air. They flew this way for several minutes before Johnson finally looked over, smiled and waved at his adversary. The German returned the salute, pulled up and headed for home.

After the German fighter left, Johnson started to climb and called "May Day" to alert search and rescue. He got an answer and a homing signal to the nearest base. It was a long way off, but he had made it this far against all odds and would make this last effort. The cockpit was full of hydraulic fluid by the time he reached Manston, so he had to make his landing without brakes or flaps. He made a perfect landing, ground-looped the aircraft to kill his speed and rolled to a stop. Bob Johnson just sat in the cockpit until the firemen got there to cut him free. His Thunderbolt was junk and would never fly again. Johnson received the Purple Heart because one of the German bullets had nicked the bridge of his nose while another had cut the band off his wristwatch.

The Group moved from the comparative luxury of Horsham Saint Faith into the mud and Nissen Huts at Halesworth on 8 July and by the end of the month they had added another three kills to the Group tally.

August of 1943 was a busy month. On 12 August the Group flew two Ramrod missions to Gelsenkirchen, using auxiliary fuel tanks for the first time. The 200 gallon tanks were made of pressed paper and mounted on the bottom of the fuselage. The tanks caused a number of problems, not the least of which was the inability to climb above 20,000 feet with the tanks attached. The Group would try a great many tank styles and sizes before they were satisfied with them and considered them combat proven. On the 17 August Ramrod missions to Regensburg and Schweinfurt, CAPT Jerry Johnson and LT Glenn Schiltz each shot down three German fighters. One more and he would become the first Ace in the Group.

Thunderbolts of the 62nd FS on the grass ramp at Halesworth. The Group moved to Halesworth in July of 1943 and it was from this base that most combat missions were flown. The P-47 in the foreground is "Kentucky Pud" flown by LT Ralph Johnson, who scored three victories. (Robert L Cavanaugh)

5

The Group almost lost Bob Johnson on 26 June 1943 when two separate battles with German Fw 190s literally shot his aircraft to pieces. The aircraft, named "Half Pint," still brought him back to Manston, England. Johnson would go on to score twenty-seven kills, but "Half Pint" was sent to the scrap heap. (USAF)

On a Ramrod to Regensburg on 17 August the Group claimed seventeen fighters shot down. Glen Schiltz got three, Walker Mahurin got two, and Jerry Johnson claimed three. Examination of gun camera films, however, revealed that both Johnson and Frank McCauley had fired on the same Fw 190, so each received half credit for the kill. On another Ramrod on 19 August, this time to Gilze Rijen Airdrome in Holland, Jerry Johnson shot down another Focke Wolf which should have been his fifth kill, making him the first Ace in the 56th. But the shared victory on the 17th meant that Johnson's total would stand at four and a half until October.

The Ramrod to Regensburg/Schweinfurt on the 17th was personally costly to one of the young pilots in the 63rd FS. COL Harold Comstock recalled the mission:

COL "Hub" Zemke, commander of the "Wolfpack," on the wing of his P-47 named "Moy Tovarish" which was painted in White on the fuselage in Russian lettering (the name was Russian for My Friend). Zemke scored sixteen air-to-air victories flying with all three squadrons before transferring to the 479th FG in August of 1944. (Don Garrett, Jr.)

The orders were quite clear and at this time, they were always the same — Stay With the Bombers! But sometimes you cannot obey the orders. I was Blue 3 and we were on our way out (escorting the bombers home). Two Me 109s came down out of the sun right in front of us with their bellies to us. A perfect angle for a kill I thought. I called them in but my leader told us to 'Let them go!' I watched them continue down and start a rollout to the left. That's when I saw that there were two Jugs right in front of them if they continued their flight path. I rolled off to the right and started down knowing full well that I would be in trouble for breaking a direct order to leave them alone and for breaking escort. As I closed on the rear Me 109, his leader opened fire on the lead P-47. He must have seen me coming as he did a quick split S and cut away. The Messerschmitt leader hit the Jug hard as I was closing on him, then he rolled right and down. I closed to about 200 yards or so and gave him several short bursts, one of which hit the left wing. A fair amount of the wing came off and the Me started a series of fast, uncontrolled rolls to the left, smoking all the way down. I pulled up and watched him go down, then called to the second Jug to join up with me since his leader had crashed. Upon my return to Halesworth I was fined 5 pounds for breaking formation. And the Group lost a good pilot.

In October of 1943 the Group's kill tally really starting to mount up. On 2 October, a Ramrod to Frankfurt produced no less than fifteen kills, fourteen by the 63rd Squadron. Included among these were triple kills by Walker Mahurin and LT Vance Ludwig. The mission was also the longest ever flown by a P-47 unit, 750 miles. On this mission COL Zemke made Ace, becoming the first Ace in the Group.

On 10 October three more Aces were crowned: Dave Schilling, Bob Johnson and Jerry Johnson. But all the victories for that month (thirty-nine) did not make up for the disaster that befell the 8th AF on the 14 October missions to Regensburg/Schweinfurt when at least sixty bombers were shot down. On 30 October COL Zemke was rotated home and COL Robert Landry assumed command of the Group.

On 5 November, LT George Hall shot down a lone Me 110, the Groups 100th victory. The 56th was the first group in the ETO to register 100 kills. On the 25th the Group tested a new concept, mass bombing by fighter aircraft. A single B-24 led the Group to the Saint Omer airdromes. Each P-47 carried one 500 pound bomb and all dropped on signal from the B-24 (which was late). Bombing results were negligible and this type of mission was not tried again until the use of a P-38 'droop snoot' was approved.

On 26 November, a Ramrod to Bremen produced a new ETO record: twenty-three kills, three probables and nine damaged. Double kills were recorded by Dave Schilling, "Gabby" Gabreski, Horace Craig, Walter Cook, Ralph Johnson and Flight Officer Velenta. "Gabby" Gabreski and Walt Cook both made Ace this day. Walker Mahurin got his second triple kill to become the first Double Ace in the ETO.

John Truluck recalled the event:

I was flying wing for Bud Mahurin. We were covering the bomber withdrawal from Bremen when we spotted a bunch of bandits trying to get into position on the bomber stream. Bud moved in behind one of the Germans and began to attack, but he was going much too fast and overshot. Prior to the mission Bud had told me that if this ever happened while I was flying his wing, I was to attack if I were in position. On this occasion, thinking that I could hit the German fighter, I chopped the throttle and pulled in behind the German. My attack was similar to Bud's and, in the process, I lost sight of Mahurin. The second element, led by Hal Comstock, with Petty on his wing, stayed with me. Bud proceeded to jump into a whole nest of Germans and destroyed three. While I was attempting to find Mahurin, I saw a German fighter on the tail of a lone P-47. When I attacked, the German broke up and turned into the sun. I followed him for

MAJ Dave Schilling in the cockpit of his first P-47, which carried the name "Hewlette-Woodmere, Long Island" and the codes LM-S. At this time there were seven kill markings under the cockpit; Schilling would eventually score twenty-three kills and command the Group after Zemke's transfer. (Bart Ingram)

about 1,500 to 2,000 feet even though I couldn't see him in the bright sun. When the German turned out of the sun there I was, sitting right on his tail. After one long burst, the German fighter exploded. I was so close that I flew through some of his debris. When I returned to base I learned that I had shot the German off COL Schilling's tail. Every time I saw COL Schilling after that, he would buy me a drink. The last time I saw him was at his house in 1950. He gave me the largest old fashioned drink that I have ever seen - a real Texas Jigger!

The Ramrod to Emden on 11 December was a really wild one. The Germans tried a new tactic: sitting high above the Group, diving down for a firing pass, then zoom-climbing back up. The idea was to disrupt the Group from its escort duties. If successful, a larger German fighter force would have butchered the bombers. But the Group kept its integrity intact and surprised the Germans in the midst of their attacks. Robert Lamb, Paul Conger and Donovan Smith each shot down three to bring the Group's score for the day to seventeen.

Donovan Smith remembered the mission:

It was a Ramrod to Emden. Some 600 B-17s and B-24s would hit the target and we would escort all the way to and from the

MAJ Loren McCollum in *LITTLE BUTCH* coded HV-W and *Francy Ann* coded HV-E about to roll at Halesford during 1943. COL Zemke called McCollum one of the best air tacticians of the war. The segmented code letters were common on 61st FS aircraft. (Fred LePage)

target. Just off the German coast we spotted a large gaggle of single engined fighters at about 35,000 feet, closing on us from the east. COL Landry had the 62nd Squadron engage the German fighters, while the other two squadrons remained as escort. The 62nd had their hands full as the Germans employed a 'dive and zoom' tactic to constantly have an altitude advantage. All they wanted to do was hit and run. Their objective was to try to get the escort to leave the bombers. Landry didn't bite. Soon after the rendezvous with the B-17s, we ran into the main German fighter force, about forty twin engined and sixty single engined fighters waiting over the coast. The twins (Ju 88s and Me 110s) were almost in position for a stern attack, while the Me 109s and FW 190s were coming from 12 o'clock. But this time we had the altitude advantage!

At about 34,000 feet Gabby (MAJ Gabreski, commander of the 61st) took us down into the German fighters. He ordered us to "let all hell break loose!" and we did. I spotted five Me 110s between 25 and 27,000 feet. I closed on the nearest three very quickly. Bob Lamb poured fire into the first one, getting hits all over the fuselage and the pilot bailed out. I lined up the middle aircraft and got off about sixty rounds, hitting the cockpit and right wing root area. Within a few seconds, the right wing just folded up and the Me 110 spun in. Following Lamb, I spotted another 110 over Emden at about 19,000. I told Aggers (F/O Aggers was my wingman) to stay close and we gave chase. Closing rapidly, the German finally spotted us and pulled up slightly, then broke into a steep dive. I led him about two rings (on the gun sight) and fired. Hits registered all over his underside and the 110 came apart. Both Aggers and I were hit by pieces of the German fighter. After the 110 blew up, I began to pull back up to altitude when something flashed past my nose. It was a Red-nosed Fw. We turned with him several times trying for a better position and I finally got a decent angle for a head-on attack. Both the German and I held our fire until the range was point blank. He missed and I didn't! The Fw passed by with his engine smoking. By the time I got turned around, the Fw was in a steep spiral, smoking badly, and the pilot finally bailed out.

1944 opened with new tactics, new propellers and another new commander. COL Hub Zemke returned from the States on 19 January 1944, seven days after the Group celebrated its first year in combat. The Group began to receive the new Curtiss Electric "paddle blade" propeller, which gave the P-47 a much better rate of climb and a higher ceiling. The new propellers meant the Group could go anywhere the Luftwaffe could — up or down.

The Group began flying split coverage for the bombers on 11 January, with half the Group covering one bomber force, while the other half covered another. They were known as 56A and 56B missions. The results of the first test were mixed. The A Group got ten kills, but seven bombers went down. The B Group had a milk run. A "big day" occurred on President Roosevelt's birthday, 30 January, when an A & B Ramrod to Brunswick resulted in sixteen kills, bringing the group total to 200.

HOLY JOE (UN-E) was the personal aircraft of CAPT Joe Egan, Jr., who scored five victories while assigned to the 63rd FS. CAPT Egan was shot down on 19 July 1944, two days after he assumed command of the squadron. (Don Garrett, Jr.)

Big Week

During February of 1944, "Pappy" Craig became the first pilot to complete a tour and rotate home. A Ramrod to Frankfurt on 11 February gave the Group its first ground kills when Zemke took two squadrons down on the deck chasing German fighters. After the fight, Zemke could not relocate the bombers and started home. As the two squadrons passed near Juvincourt Airdrome, they spotted a group of aircraft on the open ramp. Since most of the P-47s still had full loads of ammunition, Hub led a single strafing pass over the airfield. The results were two destroyed, one probable and five damaged.

But undoubtedly the highlight of the month was what has become commonly referred to as "Big Week," the period in February when the 8th Air Force set out to cripple the Luftwaffe. The bombers attacked German aircraft manufacturing and assembly centers throughout Europe, while the fighters were given free reign to attack the Luftwaffe wherever they found them, in the air or on the ground. "Big Week" began on 20 February when over 1,000 bombers attacked Leipzig, Brunswick, and Halberstadt. The 56th covered two of the bomber forces going to Leipzig. The pilots would have plenty of gas since the Group was carrying 150 gallon belly tanks for the first time. The Group shot down fourteen German fighters on the first day, including three in the traffic pattern at Steinhuder Lake. One group of nineteen Me 110s were all accounted for when MAJ Sylvester Burke led the 63rd Squadron in an attack on the rear of the German formation. "Big Week" ended on 25 February with the 56th claiming fifty-nine kills. More importantly, the Luftwaffe had been dealt a blow from which they never really recovered.

LT Wayne O'Conner and his dog pose on the wing of his 63rd Fighter Squadron P-47C. The aircraft carried a cartoon character of a fancy dressed Wolf on the nose. O'Conner scored one and a half victories. (Don Garrett, Jr.)

On 3 March, 8th AF P-38s flew over Berlin for the first time. On 4 March, 8th AF B-17s and B-24s followed the path of the P-38s — target Berlin. The 56th A and B Groups would escort. Berlin Ramrods continued on 6, 8, 9 and 22 March. The Ramrod on the 8th was notable since the Group shot down twenty-seven Germans bringing the Group total to 300. Walker Mahurin scored another triple kill, his third of the war, making him the leading Ace in the ETO and the first to score twenty victories. On 9 March, the escorted bomber force encountered no enemy aircraft all the way to Berlin and back! German fighter controllers had purposely kept the fighters on the ground.

As successful as the month was for victories (88), March ended disastrously for the Group. On an A & B Ramrod to Bordeaux and Chartres, the Group lost four pilots, including two Aces - Jerry Johnson and Walker Mahurin. Mahurin was shot down by the tail gunner of an Me 110 that he was attacking. Johnson was hit by ground fire while strafing a truck convoy. Finally, on 19 March, two returning B-24s collided over Halesworth and exploded. The Group fire fighters and medics arrived on the scene just as the bombs in one of the B-24s exploded. Six men were killed, including double ace LT Stanley Morrill and thirty-three were wounded in the mishap.

During April, the Group hit the jackpot. On 5 April the 56th conducted the first purposely flown strafing sweep of a German airfield. The code name for this mission was *Jackpot*, which was later changed to *Chattanooga*. These involved the strafing of enemy transportation targets, such as trains (ie Chattanooga Choo-Choo). The mission on the 5th was a bust due to weather over the target. But one flown on 15 April was the first time that more ground kills were scored than air kills (12-5), including a triple by COL Zemke. On the 18th the Group returned from a Berlin Ramrod to find that they were moving to Boxted near Colchester. Not only did their home change, but the missions began a subtle but steady change to that of fighter-bomber. The Group flew their first dive bombing mission against a German airfield near Paris on the 28th. It was the beginning of the buildup to D-Day.

May saw the arrival of Spring in England and the return of Bud Mahurin. Mahurin had walked out of France and was returned to the Group on 5 May after being picked up by an RAF Lysander. On 8 May, Bob Johnson got a double kill, bringing his score to twenty-seven victories, one more than the highest ranking American Ace of World War 1, CAPT Eddie Rickenbacker.

Bob Johnson remembered the events:

It was a Ramrod to Berlin. I had twenty-five kills at the time and was actually over the limit of my tour. This mission would probably be my last because the brass were not going to allow me to extend my tour. After takeoff, we rendezvoused with the bombers and began flying parallel to them. Nothing much happened. I began thinking that if I turned back now, I would land with about thirty minutes left in my tour. That's enough for at least one more mission takeoff. But fate took a hand. Just as I was about to turn the Group over and turn back, I noticed that something was happening in the middle of the bomber formation. They were under attack! It was terrible to be excited that someone else might be getting killed, but I was. I called over the RT "The bombers are getting hit! Let's go get 'em!" We turned to the

Eleanor III/El Diablo (UN-G) was the personal P-47C of LT Anthony Cavallo of the 63rd Fighter Squadron. Cavallo scored three victories with this Thunderbolt. The aircraft is armed with a 500 pound bomb on the fuselage drop tank mount. (Don Garrett, Jr.)

LT Frank Klibbe leans on the propeller of *LITTLE CHIEF/ANDERSON INDIAN* (coded HV-V). LT Klibbe scored seven victories with the 61st FS. The name "Anderson Indian" referred to the baseball team that Klibbe played for before the war. (Don Garrett, Jr.)

LT Charles Reed flew this P-47D named *Princess Pat* while attached to the 63rd FS. P-47Ds received the Curtiss Electric paddle blade propeller in January of 1944, which gave the aircraft a greater rate of climb. LT Reed scored two and a quarter victories. (Charles Reed)

left and came back on the bomber stream course. Just then an Me 109 flashed by under my nose. I immediately rolled onto my back, reversed, and tacked onto his tail. He chopped his throttle and started into a very tight left turn. I cut my throttle and rolled inside the German's flight path. We were real close. Close enough that I could see the pilot look back over his shoulder as I opened fire. He went into a dive but I kept right on his tail pouring fire into him. Suddenly his left wing came off and the fighter spun in. That made 26! We had just started back for the bombers when my #4 called out that a couple of bandits had just entered the cloud deck below us. We rolled in, catching the German fighters just as they emerged from the clouds. I lined up one of them and began shooting. He started smoking, then caught fire. I watched him roll onto his back, start a shallow dive which increased rapidly, and head straight into the ground.

The two victories that Johnson showed his crew chief on arrival at Halesworth made him the first Ace to surpass Rickenbacker's World War One record. All the brass were on hand to greet him and a big party was held that night. Within a few days Bob Johnson was on his way home.

On 10 May the Group flew their first mission from Boxted. A Ramrod to Frankfurt on 12 May raised the Group total to 400. On this mission, COL Zemke tried a new tactic which would become known as the "Zemke Fan." The Luftwaffe, using their ground radar net to both track the bomber stream and pick out any potential weak points, assembled their fighter forces near these "weak" areas to maximize their effectiveness. The Allied fighter escorts, being outside the range of RAF GCI radar, relied on eyesight to detect the waiting Luftwaffe.

CAPT Mike Quirk flew this P-47D (coded LM-K) with a Donald Duck cartoon character on the nose. CAPT Quirk scored twelve air-to-air and five ground victories while assigned to the 63rd Fighter Squadron. (Art Krieger)

COL Zemke's idea was to send out a fighter force some 30-45 minutes ahead of the bombers to scout for Luftwaffe concentration points. The object was to break up the waiting German fighter formations before they could organize any type of effective intercept of the bombers. After a number of meetings with 8th Fighter Command brass, Zemke finally prevailed and was allowed to use the tactic on the next mission.

With Zemke leading, the A Group fanned out over a semi-circle near Coblenz looking for any waiting German fighters. As it turned out, they encountered the main German fighter force, with about a dozen very skilled Me 109 pilots, who promptly shot down two of the Thunderbolts. Gabreski and B Group were nearby and rushed to the scene of the fight. Both groups found themselves in the middle of some seventy plus German fighters. The ensuing fight was intense and hectic. Everyone was engaged with a German fighter and the sky was full of wheeling airplanes and bullets. The 56th lost three pilots, but the Germans suffered the loss of eighteen plus two probables and two damaged. CAPT Robert Rankin accounted for five.

The fight almost cost the Group their famous leader. COL Zemke recalled:

Three of us had moved out to scout the Frankfurt area when about six or seven Me 109s jumped us from above. Both of my wingmen were shot down almost immediately. I went into a steep dive, spinning the airplane at the same time and the Germans broke off their attack. I headed for where I thought the rest of the Group should be but another group of four Me 109s jumped me near Wiesbaden. Again I was totally defensive, just trying to elude the German fighters. I poured on the throttle and slowly pulled away from them. When finally out of range the Germans again broke off and I started a course for England.

Somewhere near Coblenz I spotted four German aircraft about 5,000 feet below me. They were simply flying in a circular manner, probably in some kind of landing pattern. I immediately thought of jumping them from my altitude advantage, making one fast pass through them, get off some shots, then head for home. With my speed and altitude, they wouldn't even know I was there until after I was gone. I started to circle for the best position, watching the enemy formation the whole time.

LT Frank Klibbe's P-47 *LITTLE CHIEF* was repainted with a much larger Indian head on the cowling. The aircraft has the *ANDERSON INDIAN* name deleted and carries four kill markings under the windscreen. (Don Garrett, Jr.)

Bob Johnson, Hub Zemke, and Bud Mahurin stand alongside Johnson's P-47 named *Lucky* (coded HV-P). At this time Johnson had four victories which were displayed under the aircraft's name on the engine cowling in Black. (Don Garrett, Jr.)

Just as I was about to attack, I saw another gaggle of Me 109s and Fw 190s join the first four. As I sat on my perch, I watched as the strength of the German formation grew steadily until there were about thirty aircraft below me and they were slowly gaining altitude toward me. As I watched them slowly climb, I also started a slow climb. Since I knew that my group was somewhere nearby in the Coblenz-Frankfurt area, I began frantically calling for help. If they could get here before the Germans, we would still have an altitude advantage and could turn this into a turkey shoot. After about fifteen minutes, I was at 29,000 feet and making contrails. Bob Rankin and his wingman finally saw me and moved in to give me some cover. I told them to fly top cover, I was going in for the bounce. The Germans still had not seen us. I went into a fairly steep dive and picked out an Me 109 on the outer edge of the German formation. I moved into position, pulled about two rings of lead on the German and squeezed off a few rounds. They were well ahead of the Me 109 but the German flew right through them. Several hits were observed up and down his fuselage. But he did not explode.

These 63rd FS P-47Ds are undergoing an engine oil change. The P-47D was powered by a 2,300 hp Pratt & Whitney R-2800 which gave it a top speed of 426 mph. The aircraft in the foreground (UN-S) was flown by LT Sam Stamps, while UN-W was flown by CAPT John Vogt (5 victories). (Warren Bodie)

I was so close that I almost rammed him when I went by. Looking back, I watched him do a couple of quick rolls and start into a spin. His engine burst into flames, the canopy came off and the pilot bailed out. My concentration on making the kill almost cost me dearly. I had been so intent on the 109 that I had shut out the rest of the world. Rankin was frantically calling me trying to tell me that an Me 109 had tacked onto my tail. It finally came to me when I heard him yell "Break Left NOW!" I never even saw him. Rankin jumped him and shot him off my tail. But I didn't have time to think about it as I had my hands full with another four German fighters. I broke into the German formation, made a half roll, snapped off a few quick shots and again poured on the coal, running for home. When the last of the Germans had given up trying to catch me, I throttled back and started to check everything out. The aircraft seemed in good shape except that I had only 125 gallons of fuel left. And I was alone again. I cut the throttle to idle and slowly but steadily made for Boxted. I was as limp as a rag when I touched down that day.

On 19 May, the new "bubble top" P-47D-25s began arriving. These were essentially the same aircraft that the Group had been flying with one exception, a bubble canopy. This canopy gave the pilot a 360 degree unrestricted view — a great advantage when the sky is full of Germans. Glide-bombing was tried on 22 May and COL Zemke flew a P-38 "Droop Snoot" mission to the Criel railway bridge on the 30th. In the former gun bay rode "Easy" Ezzel over a Norden bombsight. The Group was loaded with 500 pound bombs, which would be dropped on Easy's signal. But the flak over Creil was too intense so Zemke took the Group five miles further south where they neatly dropped three spans of the Chantilly rail bridge.

The invasion of Europe was set for sometime in June. Everyone knew it — even the Germans. They just didn't know exactly when or where. The Groups' first indication that D-Day was at hand came when they returned from a routine patrol over the French coast and found Boxted a beehive of activity. Aircraft were being camouflaged as quickly as possible. Every man available had a paint brush and was applying large Black and White stripes on every aircraft. With paint brushes in short supply, everything thinkable was being used to apply the paint, even brooms and rags. At 1600, COL Zemke was called to 8th Fighter Command Headquarters for a special Group Commanders meeting. He returned at 2000 and immediately called a meeting of the Group intelligence officers. When they broke up, the pilots were informed to get as much sleep as possible, since the first briefing for the next days mission was set for 0130! The invasion was ON!

D-Day

As the Group prepared for their first mission in support of the Invasion, most of the aircraft were camouflaged with whatever was at hand, be it Army Air Force or RAF paints. Some were painted in an RAF style shadow camouflage using RAF Medium Sea Grey and AAF Medium Green 42. Some were given a mottle style camouflage. Some were in standard Olive Drab, while others were in Medium Green 42. Several were painted in two RAF Greys.

No matter what their camouflage style, all carried the D-Day invasion stripes which consisted of three White and two Black bands that went completely around both wings and the rear fuselage. The stripes were crudely applied without masking since they had to be finished by the time the first aircraft was ready to take off, even if they had to be finished in the dark!

The first mission was set to go at 0330, 6 June 1944. The Group flew top cover for the invasion forces, patrolling the Straits of Dover near Dungeness and Boulogne. The orders were to "be on station by 0425." The Group's mission was to watch for any torpedo aircraft attempting to attack the troop ships. The first two aircraft rolled down the Boxted runway at 0325. Two by two, the 61st and 63rd Squadrons took to the air. At 0608 the 62nd Squadron left Boxted to relieve the first two squadrons over the Channel.

The Germans were still unsure if this was the invasion and the Luftwaffe was slow to react. The first two missions were uneventful and the third and fourth missions were fighter bomber in support of the GIs. On the fifth mission the Luftwaffe came up ready to fight. LT Wm. McElhare, a former B-17 pilot, got an Fw kill when the German tried a reversal and snap-rolled into the ground. On the last mission of the day, COL Zemke led the 61st and 62nd Squadrons on a fighter sweep to Rouen. As they were ready to return home, Zemke spotted some Mustangs in a fight with about fifteen Fw 190s and decided to join in.

When I was almost in position to bounce some Germans, I saw a single Fw diving on two Thunderbolts. I turned right and started down. He must have seen me, since he broke off his attack and dove to the west. With an altitude and speed advantage I began to overtake him. He turned abruptly to the right and attempted to engage me. After a tight 180 turn, the Fw tried to reverse, lost control and spun in. I used the gun camera to record the fireball on the ground.

At 2218 the last P-47 landed at Boxted; D-Day had ended.

Bob Johnson congratulates CAPT Walker "Bud" Mahurin on his fourteenth victory, while the crew chief paints the kill marking under the canopy. Mahurin's aircraft (UN-M) was a "presentation aircraft" paid for by the residents of Atlantic City, New Jersey. Mahurin scored seven more victories with the 63rd FS. (USAF via Jeff Ethell)

The large hole in the fuselage above the White "V" on LT Praeger Neyland's *Pistol Packin' Mama* (coded HV-B) was caused by a 20mm shell from an Me 109. Although the German fighter knocked out Neyland's radio, oxygen system and tailwheel, he still brought the P-47D home. (USAF)

MAJ Sylvester Burke congratulates "Gabby" Gabreski on the occasion of his birthday, his 11th victory and his promotion to LTC, all on 25 January 1944. Gabby would score seventeen more kills before making a crash landing and being captured during the Summer of 1944. (USAF)

LT Albert Knafelz of the 62nd FS had the cowling of his P-47D (LM-A) Thunderbolt painted with Donald Duck imprisoned in *Stalag Luft III* even though he *Wanted Wings*. (Art Krieger)

LTC Francis "Gabby" Gabreski flew this Thunderbolt coded HV-A. None of his aircraft carried names but each was distinctive. This aircraft has one of the shadow-shade camouflage schemes applied for the D-Day invasion. The aircraft has AAF Medium Green 42 and RAF Medium Sea Grey uppersurfaces over Medium Sea Grey undersurfaces. (Imperial War Museum)

LTC "Gabby" Gabreski and his ground crew, SSGT Ralph Safford (left) and CPL Felix Schacki (right), on the occasion of his 27th victory. This kill tied him with Bob Johnson as the leading Ace in the ETO. Gabby would add one more victory before becoming a POW. (56 TTW/HO)

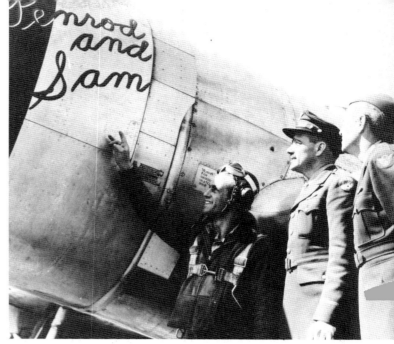

Bob Johnson is congratulated by BGEN Bill Kepner, Commanding General of 8th Fighter Command along side *Penrod And Sam* (coded LM-Q). This was the last Thunderbolt Johnson flew while assigned to the 62nd FS. The aircraft was named for Johnson's crew chief, Penrod, and Johnson's middle name, Sam. (Don Garrett, Jr.)

Oregon's Brittania (coded Z) was flown by COL Zemke. The P-47D, carried a camouflage of RAF Medium Sea Grey and AAF Medium Green 42. The Black and White D-Day bands were applied very crudely, usually without masking. (Robert Elliot via Robert Cavanaugh)

Hit by flak over Normandy, this Wolfpack P-47D made it back to England and crash landed in a field. The pilot was unhurt but the aircraft was scrap and was stripped of any reusable parts to help keep other P-47 operational. (David Menard)

CAPT James McClure was an expert at most anything, including the unicycle. The Thunderbolt in the background is LT A.P. Knafelz's *Stalag Luft III/I Wanted Wings*. The dark area over the national insignia was where the D-Day stripes were painted over. (USAF)

By the end of the day, the total sorties flown by 8th Fighter Command had reached 1,813. The 56th FG had flown 142 of those. Most of the missions were in support of the ground troops. While one squadron blasted a path for the GIs, the other squadrons flew top cover. Total kills for the day was twenty-six (three by the 56th). 8th AF losses were twenty-two P-51s and four P-47s (one from the 56th). Most of these losses were from ground fire. The rest of June was a repeat of D-Day, ground support missions. On the last day of June, a Boxted pilot became the first 8th Fighter Command pilot to shoot down a German V-1 Buzz Bomb when LT J. Tucker from the 5th Air Rescue Squadron splashed one into the Channel.

In July the Group was back flying air superiority missions. Gabreski became the leading Ace in Europe on 5 July when he scored his 28th kill. On 7 July Fred Christensen put his flight into the traffic pattern over Gardelegen Airfield with twelve German Ju 52 transports. Christensen shot down six, Billy Edens got three and Mike Jackson got one. July was also costly as Joe Egan, commander of the 63rd Squadron, was killed on the 19 July Ramrod to Augsburg and on the 20th, Gabreski was shot down and captured. It was later learned that Gabby had gotten too low while strafing Bassinheim Airfield and caught his prop on a small hill.

On 17 August, Operation MARKET GARDEN, the Allied airborne invasion of Holland, began. The Group's mission was flak suppression, considered to be the most hazardous mission of all. It turned out to be true. On the second day of the operation, the Group lost sixteen Thunderbolts, with another twelve damaged. Four pilots were killed, one was shot down and captured, three were badly wounded, and eight others either bailed out or bellied in somewhere in Allied territory. It was the highest loss total for the Group at any time in the war.

COL Harold Comstock told the story of the operation:

I was awakened by Schilling and told that I would lead the Group. I picked up the Operations Order, checked the weather, and went back to the "Wheel House." I told Dave that the weather was very bad and we should try and get out of the mission. But the order read that we would go at "all cost," that we were not to shoot until shot at and that the mission was flak suppression ahead of the supply aircraft at Arnhem. Target area weather was about 500 feet, with a top layer about 2,000 feet. The bomber lead reported that he was under the clouds so I spread the three squadrons out and had them drop down to that altitude. At that point my wingman went down and aircraft were being hit all over the place. The bombers were taking heavy fire and it was obvious there was nothing we could do to help. It was all over in about five minutes. A Lieutenant from the 61st and I made rendezvous, where I told the rest of the guys to break it off. The LT and I watched one of the B-24s fly smack into a brick smoke stack and blow up. There was all kinds of battle damage and Dave asked me why I didn't abort the mission. I reminded him of the "at all cost" Operations Order and our earlier discussion. The Group was awarded a Presidential Unit Citation for the mission.

CAPT Donovan Smith's *OLE COCK III* (coded HV-S) parked on the Boxted ramp during the late Summer of 1944. The aircraft is armed with 4.5 inch bazooka rocket tubes under each wing. A fuselage drop tank was usually carried in conjunction with the bazooka tubes. Donovan Smith scored 5.5 victories while in the 61st FS. (Robert L. Cavanaugh)

Pat a P-47D of the 63rd FS flies over the Channel enroute to Boxted after an escort mission during the Summer of 1944. By July of 1944, the uppersurface D-Day stripes had been removed or, as in this case, painted over. (AFM)

This P-47D (coded LM-F) of the 62nd Squadron was one of a number of Thunderbolts used by LT Billy Edens. Although ten kills are painted under the canopy, Edens was "officially" credited with seven victories. The P-47 was damaged when it collided with a P-51 on landing. (Don Garrett, Jr.)

On 21 August the Group, or what was left of it, escorted the airborne landings at Arnhem. The Germans were waiting! Two squadrons of fighters jumped all over the Group, penetrating through to the transport force. Although the Group shot down fifteen German fighters, the transports were sitting ducks. Losses included at least thirty-five C-47s and fourteen Stirlings. The Luftwaffe was definitely not dead yet!

During August, 8th AF P-51s escorted a bomber force all the way to Gdynia, Poland, on the first shuttle mission. Paris fell to Allied units and COL Hub Zemke, the Group's leader from the very beginning, was transferred to the 479th Fighter Group, a P-38 outfit! COL Dave Schilling took command of the 56th on 12 August. The total kills for the month were aircraft 20 kills/11 damaged, trucks 145/74, locomotives 88/78, rail cars 324/912, wagons 45/9.

September saw the Group set another single day (5 September) record, seventy-eight confirmed, including sixty on the second mission, a strafing sweep of Gelnhausen Airfield.

On 1 November the Group scored its first kill against the new German jet fighters. It took two groups of P-51s plus the 56th Thunderbolts to bring down the speedy German interceptor. For most of the war, the pilots of VIII FC were usually outnumbered by the Luftwaffe. But on 1 November a lone German fighter pilot challenged not only an entire bomber formation, but also the three groups of escorting American fighters. And he almost won! The German was flying a Messerschmitt Me 262 jet fighter, it was some 150 mph faster than any Allied fighter aircraft and carried four 30MM cannons, one hit from these could knock down a bomber.

The Group was covering the bomber withdrawal over Holland when MAJ Harold "Bunny" Comstock sighted a jet contrail at about 38,000 feet. The jet went into a shallow dive, pulled in behind one of the P-51s and exploded it. The Mustangs from the 352nd and 20th FGs, joined the 56th Thunderbolts in the chase. The Me 262 made a diving left turn, leveled off at about 10,000 feet, then turned hard

This P-47D-25 Thunderbolt (coded LM-P) has a bubble canopy and was flown by Lucian Dade during June of 1944. Dade scored six victories and commanded the Group during 1945. (Robert L Cavanaugh)

right and accelerated away toward the Zuider Zee.

The German pilot had to be slightly overconfident in the performance of his jet with his easy kill over the Mustang. Instead of heading for home to fight another day, he turned back toward the three groups of American fighters. The Americans cut off the German jet before it could break away. Both the P-47s and the P-51s turned into the jet's flight path, with at least six pilots taking a crack at it, including LT Walter Groce of the 63rd FS. The jet began a climbing turn to the right and Groce closed in. Both Groce and LT W. Gerbe from the 352nd had the right angle and began firing. Groce pulled his nose through the jet's fuselage, spraying him with bullets. The right engine blew up, the Me 262 went into a spin and the pilot bailed out. Groce and Gerbe shared in the kill.

In December of 1944 the Germans launched what has become known as the Battle Of The Bulge. They timed it perfectly as bad weather socked in all the Allied airfields for days allowing the German armor to operate in daylight, something they would never have considered in clear weather. If the weather had held, the Germans might have succeeded in driving all the way to Arnhem and splitting the Allied forces in two. But GEN George Patton's Third Army stopped the Germans at Bastogne and the weather broke just before Christmas.

About 1140 on 23 December, COL Dave Schilling was leading the Group on a sweep in the Bonn area when he ran into over 250 German fighters. GCI radar had vectored the Group toward a large formation of bandits, but they lost them in the cloud cover. Returning to their original heading, Schilling sighted another large enemy formation, again losing them in the clouds. Irritated that GCI had not notified them of the new target, Schilling called the controller for some answers.

GCI came back with "Don't worry about it! There's bigger game on this heading!" Bunny Comstock was the first to see them, calling out a large formation of Fw 190s below them. At the same time Schilling spotted another group of 40+ about 1,500 feet below and slightly ahead of him. He ordered the 61st and 63rd Squadrons to handle those directly below, while he took the 62nd Squadron against the other German squadrons.

Pat, a P-47D of the 61st FS, flies wing on MAJ James Stewart in his Thunderbolt coded HV-K. *Pat* was originally a Natural Metal aircraft which was camouflaged for D-Day operations. The D-Day stripes have been painted over. (Robert L Cavanaugh)

62nd FS ground crewmen fill the main fuselage tank of this P-47D Thunderbolt. With underwing drop tanks, the P-47D had a range of over 1,000 miles. The aircraft is equipped with a Curtiss Electric paddle blade propeller. (Art Krieger)

Ground crews pose on *Silver Lady* during the Summer of 1944. The aircraft was flown by FLT LT Witold Lanouski, one of six Polish Air Force pilots attached to the 56th FG. Lanouski scored six victories, two with the Royal Air Force and four with the 61st FS. (Merle Olmsted)

MAJ Harold Comstock leads a flight of four P-47D Thunderbolts of the 63rd FS back from another Rodeo mission over Germany. Pilots from the 63rd FS accounted for 267 1/2 German aircraft destroyed, with MAJ Comstock scoring five confirmed victories. (AFM)

LTC David Schilling in *HAIRLESS JOE* (coded LM-S) and LT George Bostwick in *Ugly Duckling* (LM-Z) begin their takeoff roll for another mission in the Summer of 1944. Schilling and Bostwick were both Aces with 23 and 8 victories respectively. (Robert L. Cavanaugh)

The Germans were very aggressive in their new, long-nosed Fw 190Ds. But the element of surprise and the aggressiveness of the 56th pilots took its toll. Comstock and seven others got two apiece. Schilling had his squadron simulate a German formation and pulled right in behind the unsuspecting fighters. The 62nd was all over the Germans before they even knew they were there. Schilling shot down three Me 109s, later adding two Fw 190s for a total of five confirmed. The total for the day was another record: thirty-seven destroyed, one probable and sixteen damaged. This brought the Group total to over 800 destroyed, or roughly 25 percent of all the Germans shot down by 8th AF fighters!

By January of 1945 the 56th FG was the only 8th AF fighter group not flying the P-51 Mustang. The P-51 was faster, flew higher and had a greater range. But the Group stubbornly clung to their tough old Thunderbolts. Republic was about to introduce a new, higher performance variant, the P-47M. The P-47M was Republic's answer to obtaining maximum performance from the standard P-47D air-

frame. A new engine, the P&W R-2800-57 rated at 2,800 hp in War Emergency setting, coupled with a new Curtiss Electric propeller raised the top speed to over 470 mph. The only group in the Air Force to get these new Thunderbolts was the 56th.

The first P-47M arrived at Boxted on 3 January 1945. But the P-47M had problems that were not fully ironed out by the time the aircraft reached the combat zone. The higher speed brought some instability problems that were ironed out with the addition of a fin fillet. Problems with engine cylinder head temperatures and ignition breakdown at altitude were harder to solve. Finally all the engines were removed and replaced with new engines which cured most of the problems. The Group took the first P-47M into combat on 4 January in a fighter sweep to Magdeburg. But the Group was in a non-combat status for almost a month trying to solve all the mechanical problems. On the 22nd another "old hand" was transferred out of the Group when COL Dave Schilling moved up to the 65th Wing Headquarters. COL Lucien Dade took over the Group.

By February, enough P-47Ms had been delivered to equip all

CAPT Fred Christensen leans on the prop of *Miss Fire*. Christensen scored twenty-two victories with the 62nd FS. Like many other "old hands," Christensen's aircraft code letter remained the same on all of his assigned aircraft. (AFM)

P-47Ds and P-51Ds at Bottisham during the August of 1944 Commander's Meeting. *GENTLE ANNIE* was flown by COL Harold Rau (20th FG); *Straw Boss 2* was LTC James Mayden's (352nd FG) Mustang; *Hairless Joe* COL Dave Schilling's Thunderbolt (56th FG) and *DA QUAKE* was flown by COL John McGinn CO (55th FG). The others are "Judy" flown by COL Phil Tukey (356th FG) and LH-E, COL Ben Rimerman's P-47D. The P-38J on the right was flown by COL Zemke, now CO of the 479th FG. (USAF)

MAJ Harold Comstock, commander of the 63rd FS, leads the squadron back to Boxted during the Summer of 1944, in *HAPPY WARRIOR*. Comstock's aircraft was previously COL Zemke's aircraft. There are two aircraft with the same code, UN-R. The camouflaged aircraft had a bar under the code indicating it was the second aircraft in the squadron with the code letters. (Robert L Cavanaugh)

1st LT Claude Chinn's UN-C on the ramp at Boxted during late 1944. LT Chinn scored seven ground victories while assigned to the 63rd FS. (Don Garrett, Jr.)

three squadrons. With the longer range of the P-47M and Allied airfields on the continent, the first planned mission to Berlin by P-47s was scheduled for 3 February. Now the Group could go to Berlin and engage the Reich Defense Forces, without worrying about fuel. Sweeping out ahead of the bomber force, the Group jumped a formation of fifteen Fws near Berlin, shooting down nine. On the 26th the Group escorted the bombers to Berlin with "no enemy aircraft reaction." The Luftwaffe was finally dying.

During February the Group began applying gaudy paint schemes to their aircraft. Both P-47Ds and Ms were camouflaged in colors that certainly made no sense to either the RAF or the Germans. The air and ground crews were quite proud of the Group's position as leaders for the honor of being the highest scoring outfit in Europe. They wanted their aircraft to really stand out. Each squadron adopted its own "camouflage" colors and scheme. The 61st Squadron painted their P-47s with Midnight Blue uppersurfaces. The 62nd Squadron shadow-shaded the uppersurfaces in Dark Green and Medium Sea Grey. The 63rd Squadron camouflaged their aircraft in Dark French Blue and Light Azure Blue. All undersurfaces remained in Natural Metal. The identification codes on the fuselage were also significant: Red and White for the 61st, Yellow on the 62nd, and Natural metal on the 63rd aircraft. The Red cowl ring, identifying the 56th FG, was retained while rudders and serial numbers were painted in squadron colors: Red (61st), Yellow (62nd) and Blue (63rd).

Lorene (HV-J) of the 61st FS was LT Russ Kyler's P-47M Thunderbolt during 1945. LT Kyler was credited with three air and seven ground victories. Rather than repaint the nose art on the P-47M, Kyler's crew chief simply removed the cowling from the older P-47D and put it on the P-47M. (Don Garrett, Jr.)

The Brat was flown by LT Randall "Pat" Murphy, a double Ace with ten ground kills on one day alone and two air victories. The Brat was one of the new P-47Ms painted in a two-tone Blue camouflage. The code letters of camouflaged 63rd FS P-47Ms were in Natural Metal. (Warren Bodie)

Flight Lieutenant Witold Lanouski was one of the Polish Air Force contingent assigned to the Group. He flew two tours and shot down three Fw 190s and one Me 109. Lanouski's P-47M (HV-Z), carried nose art showing a fist smashing an Me 109 over the Polish national insignia. The aircraft was camouflaged Flat Black, with Red code letters and a Red rudder. (AFM)

LM-L was one of the aircraft used by MAJ Leslie Smith during late 1944. The D-Day stripes have all been removed except for under the fuselage. MAJ Smith scored seven victories and became commander of the 62nd FS during September of 1944. (Tom Ivie)

Sugie (UN-T) was a P-47D assigned to the 63rd FS P-47D during the Fall of 1944. The Thunderbolt arrived at Boxted in Natural Metal, was camouflaged and given D-Day stripes, then stripped of her camouflage again within a few months. Camouflage paint added about 40-50 pounds to the aircraft and cost some 7-10 mph in speed. (Don Garrett, Jr.)

19

Category "E" was a "war weary" P-47D that was converted into a two-seat "DoubleBolt" trainer. The entire conversion was done at Boxted by SGTS Thurman Schreel and Charles Taylor. The two seater was used as both a unit hack and as a VIP transport. It was camouflaged two-tone Blue similar to other 63rd FS aircraft. (USAF)

LT Philip Kuhn leans on the drop tank of *Fire Ball* (UN-K), the P-47M that he flew with the 63rd FS. Almost the entire P-47M production run went to the 56th FG and they were the only unit in the Army Air Force to fly this variant of the Thunderbolt. (AFM)

The aircraft that the Group sent to Paris for the Victory display was the P-47M assigned to LT Paul Dawson. The aircraft was specially marked to proudly display the Group's wartime record and COL Zemke's personal UN-Z identification code.

In March of 1945 the American First Army captured the Ludendorff Railway Bridge at Remagen, and American troops crossed the Rhine into Germany. The Germans frantically tried to destroy the bridge with what was left of their air power. But 8th Fighter Command ordered an aerial umbrella over the bridge during daylight. Every conceivable German aircraft type flew against the bridge, from Stukas to jets, fighters and bombers. The Group took its turn protecting the bridge on 10 March, encountering six Messerschmitt 109s carrying bombs bound for the bridge. Upon seeing the waiting Thunderbolts, the Germans promptly jettisoned their bombs in a nearby field and fled, with 56th pilots shooting down two of their number.

German jets had become an increasing menace to the "Big Friends" once Hitler had been talked out of using the Me 262 as a bomber. The swift twin-jet fighter began terrorizing the bomber streams and there was little that the Allied fighter pilots could do if the jet was "at speed." The top of the line propeller fighters were still

MAJ Paul Conger's armorer was CPL Milhorn. He was responsible for the care of the Thunderbolt's eight .50 caliber machine guns on his P-47M named *Bernyce*. MAJ Conger was credited with eleven and a half victories flying with the 61st and 63rd Squadrons. (Robert L Cavanaugh)

LTC Dade poses beside the Paris Display Thunderbolt which carried the name *Zemkes Wolfpack* and the inscription *OVER 1000 E. A. Destroyed* on the cowling. (AFM)

some 150 mph slower than the German jet. Luckily, the decision to use the Me-262 as an interceptor came too late to effect the outcome of the war.

The Group did manage several air-to-air Me 262 kills with the P-47Ms. In one engagement, the Group met some fourteen Me 262s near Oranienburg. Four were at about 25,000 feet, with the others below and behind. The added speed of the P-47Ms caught the Germans by surprise and two of the German jets were shot down. The others fled for home. On the way home, the Group strafed several German fighter bases resulting in an additional forty-five and a half destroyed. This brought the Group total to over 900.

On 13 April 1943, the Group flew its first combat mission. They celebrated this event two years later with a "Free Lance" mission into Germany. Arriving over Eggebeck Airfield, the Group found over 150 German aircraft sitting on the ramp. COL Dade led the Group down on the deck and turned them loose. Every pilot used almost all of his ammunition. The result was an incredible ninety-five destroyed and ninety-five damaged. The day's total took the Group's wartime claims to over 1,000 Destroyed.

On 21 April 1945, the A Group flew a Ramrod mission to Salzburg, while B Group conducted a Free Lance in the Munich area. When the bombers aborted, the A Group joined up with B and went hunting. They found nothing worthwhile and returned to Boxted. The mission was recorded as "Uneventful." However, it would become the most eventful mission of them all — the *last one*.

In the two years and twelve days that the Group was in combat, they had flown 447 missions, 19,391 sorties and 64,302 combat hours. They had destroyed 665 ½ German aircraft in the air and 324 on the ground; probables amounted to 58 and they damaged a further 243 in the air and 300 on the ground. A total of 1,590 ½ German aircraft were put out of action. In accomplishing this they had used 3,063,345 rounds of .50 caliber ammunition. The Group had forty-nine Aces among its pilots, including the top two in Europe: Gabby Gabreski (28) and Bob Johnson (27).

Although it appears to be in overall Light Gray, this P-47D flown by LT Claude Chinn of the 63rd FS was camouglaged in Light Sea Gray and Dark Sea Gray uppersurfaces with a Medium Blue rudder. (Charles McBath via James Crow)

This He-111 (coded HV-COW) was "captured" by MAJs Felix Williamson, James Carter and John Ordway and used as a group hack during the Summer of 1945. The aircraft was Flat Black with Red cowlings, rudder and codes with White trim. (COL James Carter)

This captured Fw 190 was flown to Boxted for tests against the Group's P-47Ms. It was later found that the engine had been sabotaged and the aircraft was scrapped after its systems were carefully studied. (COL James Carter)

Jets

With VE Day behind them, the men of the 56th FG began to relax and think of going home. The non-flying personnel, who had performed so brilliantly in keeping the fighters flying, were given aerial tours of the Continent they had helped liberate. The "Big Friends" came to the fighter bases and picked up the ground crews, bringing them home seven hours later after visiting the places that only the pilots had talked about — Berlin, Dusseldorf, Frankfurt and of course Paris. It was quite a sight from the air.

During late May of 1945, the Group was notified that they would be part of the Occupation Force in Germany, based at Gablingen near Augsburg. The "high point men" (those with the greatest number of combat hours or service time) would go home; the low time personnel would go to Germany or to the Pacific.

These plans were drastically altered in June when they were alerted for duty in the Pacific and Far East training was begun. The Group would not, however, be taking their P-47s with them; they were scheduled to transition to the P-51 Mustang. Once again these plans were changed with the bombing of Hiroshima and VJ Day. During August, COL Dade was transferred to Strategic Air Force Hqtrs in Paris and LTC Donald Renwick took over the Group. On 15 October 1945, what remained of the Group's personnel boarded the

This overall Black Douglas WB-26C Invader was used by the Group at Selfridge as a unit hack and weather ship during 1946. The aircraft in the background are P-47Ns and P-51Hs. (Warren Bodie)

The 56th FG was deactivated during October of 1945, then activated in May of 1946 at Selfridge Field, Michigan. They were equipped with North American P-51H Mustangs. This P-51H from the 61st FS had a Red spinner, Yellow vertical fin, and Blue and White striped rudder. (David Menard)

QUEEN MARY at Southampton and headed for home. The Group was deactivated at Camp Kilmer, New Jersey on 18 October 1945.

Seven months later, on 1 May 1946, the Group was reactivated at Selfridge Field, near Detroit. Many of the Group's original personnel that were still in the Army Air Force were found and reassigned to the 56th. The Group Commander was an old friend, LTC David Schilling. They were assigned P-51H Mustangs and a number of their "old friends" — P-47N Thunderbolts. The P-51H and P-47N reflected their new mission: long range fighter escort of Strategic Air Command bombers.

The first jet fighters arrived in April of 1947, when the first Lockheed P-80A Shooting Stars landed at Selfridge. The Group quickly transitioned to the Shooting Stars and COL Schilling, using his wartime experience as a leader and teacher, began to bring this new group of pilots and aircraft to combat ready status. Although there was no shooting war in progress, rumblings from abroad had begun and something called the "Cold War" had started. Group training in cross country, long range flying was a daily experience. Their mission with SAC was long range escort and COL Schilling decided to test his new crews and aircraft by making a round-trip flight to England (later extended to Furstenfeldbruck, Germany).

Although the P-51H was the primary aircraft assigned and used by the 56th FG at Selfridge, several P-47s found their way onto the Group's inventory. The Group operated an all-Thunderbolt aerobatic team. This P-47N had a Blue cowl flash, Red vertical tail with Yellow serial (488680) and a Red, White, and Blue striped rudder. (David Menard)

Ah'm Available was a P-51H Mustang of the 62nd FS on patrol over Alaska. The 62nd FS deployed to Ladd Field, Alaska, during 1946 for cold weather training. The propeller spinner, wing bands and tail band are in Yellow with Black trim. (USAF)

The exercise was called *Fox Able One*, the first trans-Atlantic crossing by a major USAF jet combat group. The goal was to show the world, and the Soviets in particular, that the US Air Force could deploy a major jet force anywhere in the world within a few days of the outbreak of a crisis. Schilling didn't know it at the time, but the first real Cold War crisis was about to start. Just a few weeks prior to the planned departure date, the Soviets blockaded the city of Berlin. All U.S. armed forces were put on alert.

Fox Able One would be accomplished in seven steps: Selfridge to Bangor, Maine; then to Goose Bay, Labrador; to Bluie West One on Greenland; to Meeks Field, Iceland; to RAF Stornoway, Scotland; to RAF Odiham, England; and finally Germany. Sixteen F-80Bs departed Selfridge on 12 July 1948 and among the pilots were some well known veterans of the Second World War. Schilling, Donovan Smith, Frank Klibbe, Ron Westfall, and Clay Tice. The Group would be escorted by Air Rescue SB-17 Dumbos, with constant weather updates from a SAC WB-29 weather ship. Support personnel and equipment would make the crossing in MATS C-54s and C-47s.

The total flight time for the crossing was approximately fourteen hours, but it took the Group thirteen days to make the flight. The culprit was weather, either at the takeoff field or at the next destination. At every stop, the Group was hailed with warm greetings and welcoming ceremonies. The Group arrived intact at Fursty on 25 July after a flight of 4,740 miles. They were welcomed with open arms by the German people and press; however, the German ambassador made an official objection to COL Schilling's display of his entire Second World War "scoreboard" showing his twenty-three German kills on his F-80. Politics prevailed and the scoreboard was removed.

The 56th FG joined the jet age during April of 1947 when they began to transition to the Lockheed P-80A Shooting Stars. This P-80A of the 61st FS was painted in Gloss Light Gray with a Red band on the vertical tail and a Red arrow on the tip tanks. (David Menard)

Jackie, a P-51H of the 62nd FS, on the Selfridge ramp during the Summer of 1947 after the squadron returned from Alaska. During this time period, the Red bar was added to the national insignia. (Art Krieger)

For the next two weeks the Group visited U.S. installations throughout Germany. Old target areas and former enemy air bases were buzzed at treetop level. The Group flew a tight formation with forty-eight P-47Ds of the 86th FG at Neubiburg in a tribute to the troops on occupation duty. The Group departed Germany on 14 August, with all sixteen F-80s arriving at Selfridge on 21 August. The success of the mission led to the use of the first ferry flights to re-equip our units in Germany with jet aircraft.

On 1 December 1948, the mission of the Group was changed from escort to air defense and the 56th was assigned to the Air Defense Command, under the operational control of Continental Air Command. Operation *Fox Able Two* took place when LTC Irwin Dregne led fifteen F-80s and TF-80s across the Atlantic to re-equip one of the units still in Germany. The 56th swept the Gunnery Meet

An F-80A Shooting Star of the 61st FS starts its J-33 engine to begin the first all-jet unit crossing of the Atlantic Ocean. Operation *Fox Able One* started at Selfridge on 20 July 1948 and ended at Furstenfeldbruck, Germany, on 1 August 1948. (AFM)

Originally designated the TF-80C, the Lockheed T-33 was the trainer version of the F-80 fighter and every jet squadron had at least one assigned to it for proficiency training. This T-33A was assigned to the 62nd FS with a Yellow band on the tail and a Black border and number. The aircraft carries two .50 caliber machine guns in the nose. (W. J. Balogh)

F-80B Shooting Stars of the 62nd FS on the ramp during 1948. The Buzz Numbers reflect the change in designation from P for Pursuit to F for Fighter aircraft. The nose, tip tanks, and tail are Red and Yellow with Black trim. (W. J. Balogh)

at Las Vegas AFB that same year; winning aerial and ground gunnery, as well as the dive and skip bombing competitions.

On 20 January 1950, the Group was redesignated a Fighter Interceptor Group. Under ADC/ConAC regulations, the Group was required to maintain a constant 24 hour alert posture. The Group's area of responsibility made it part of the Eastern Air Defense Force. On 25 April 1950, the 56th FIG began conversion to the North American F-86 Sabre and by August, they were completely equipped with F-86As. All the F-80s were gone, except for a pair of TF-80Cs (T-33As) which were assigned to each squadron for training duties.

In June of 1950, several changes occurred that came quite unexpectedly and fast. The North Koreans invaded South Korea on 25 June, beginning three long years of war. Although the 56th FIG did not take an active combat role in Korea, the war did cause a number of changes within the Group. First ADC was discontinued on 1 July 1950 with the Group now under Continental Air Command (ConAC) control. Then, with the threat of a Soviet atomic attack from over the North Pole, the 62d FIS moved to Orchard Place Air Force Base (later O'Hare International Airport) in Chicago on 4 August. The 61st FIS moved to Oscoda, Michigan in early October. In December of 1950, the Group gave up its best F-86As to the 4th FIG, which took them to Korea.

On 1 January 1951, the USAF re-established Air Defense Command and assigned the 56th FIG to it. Because of a shortage of F-86s, the 61st FIS returned to Selfridge and re-equipped with the Lockheed F-94B Starfire, the first unit in the Air Force to operate the Star-

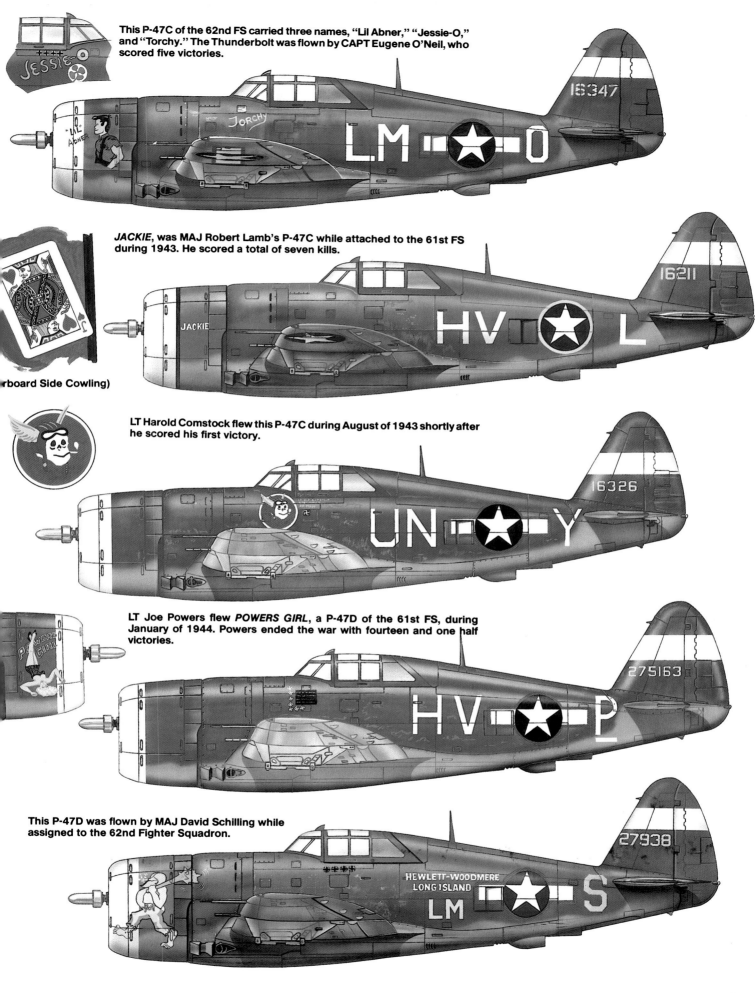

This P-47C of the 62nd FS carried three names, "Lil Abner," "Jessie-O," and "Torchy." The Thunderbolt was flown by CAPT Eugene O'Neil, who scored five victories.

JACKIE, was MAJ Robert Lamb's P-47C while attached to the 61st FS during 1943. He scored a total of seven kills.

(Starboard Side Cowling)

LT Harold Comstock flew this P-47C during August of 1943 shortly after he scored his first victory.

LT Joe Powers flew *POWERS GIRL*, a P-47D of the 61st FS, during January of 1944. Powers ended the war with fourteen and one half victories.

This P-47D was flown by MAJ David Schilling while assigned to the 62nd Fighter Squadron.

The 56th FG participated in the 1948 National Air Races in Cleveland by sending an air demonstration team. This F-80A was the personal aircraft of LTC W. Ritchie, commander of the 61st FS. The aircraft had the flags of all the nations the Group visited during the cross Atlantic mission painted on the fuselage. (Peter Bowers)

The Operation *Fox Able One* return flight stopped for refueling at Dow Air Force Base, Maine. The aircraft in the background are Republic F-84B Thunderjets assigned to the 14th Fighter Group. (USAF via David Menard)

fire. On 5 January 1951, the 63rd FIS took their F-86As to Oscoda (now known as Wurtsmith AFB). Finally, in May of 1951 the Group was reinforced with the addition of two Air National Guard squadrons, the 172nd FIS Michigan Air Guard and the 136th FS New York Air Guard. The 172nd FIS brought F-51Ds with them to Selfridge and the 136th FS was flying F-47 Thunderbolts, the last P-47s to be assigned to the Group. During this period, many of the Group pilots were rotated through Korean combat units, including Gabby Gabreski who became a Jet Ace while commanding the 51st FIW in Korea.

On 6 February 1952, under a major Air Defense Command reorganization, the 56th FIG was inactivated. Its squadrons remained in place but under the control of other Air Defense Wings: the 62d FIS at Chicago under the 4706th ADW; and the 61st at Selfridge and 63rd at Oscoda under the 4708th ADW. This changed again on 16 February 1953 when the squadrons were assigned to groups within the Air Defense Wings: the 61st to the 575th ADG, 62d to the 501st ADG, and 63rd to the 527th ADG.

Modernization of the air defense forces began during late 1952 when the 63rd FIS received F-86E and F Sabres. The 62d FIS transitioned to F-86D Sabres in March of 1953 and finally, in July of 1953, the 61st FIS, now assigned to North East Air Command, took their F-94Bs to Ernest Harmon Air Base in Newfoundland. Here they transitioned to Northrop F-89D Scorpions during the Spring of 1954. By January 1955, the 63rd FIS had turned in its F-86E/Fs for F-86D Sabre Dogs, then transitioned to F-89D Scorpions.

The 56th FG swept the honors at the 1949 USAF Gunnery Meet at Las Vegas AFB, Nevada (now Nellis AFB). The Group took the honors in air and ground gunnery and swept the bombing competition. This 63rd FS F-80B has a Red nose and a Blue band on the tail. (USAF)

During January of 1950, the 56th Fighter Group became the 56th Fighter Interceptor Group and began conversion (in April) to the North American F-86A Sabre. Part of their new mission was a 24 hour alert, which meant night refuelings such as this one. (USAF)

On 18 August 1955, the Air Force initiated Project *Arrow*, which brought back historical group designations and resulted in the renaming of many of the Groups that had been changed during 1952. The 56th Fighter Group (Air Defense), with headquarters at Old Orchard Field, Chicago, was re-activated within the 4706th ADW. The 62nd and 63rd FIS 'designations' moved with the Group Headquarters, transitioning back to F-86D aircraft. During October of 1957, the 61st FIS transferred from Harmon Field to Truax Field, Wisconsin under the control of the 327th FG (AD) and went supersonic when they transitioned to the Convair F-102A Delta Dagger interceptors.

The Group slowly began to dissolve early in 1958 when the 63rd FIS was inactivated on 8 January. The 62d FIS went to K.I. Sawyer AFB, Michigan during August of 1959 and was assigned to the 473rd FG (AD). This squadron, although not assigned to the 56th FG (AD) was the first to have a "double-sonic" capability as they turned in their aging F-86L Sabres for McDonnell F-101B Voodoo's. The 61st FIS with F-102As was still attached to the 327th FG(AD).

COL Robert Casey of the 62nd FIS flew this F-86A named *Texas Tailpipe*. The aircraft is parked inside the alert barn at Orchard Place Air Force Base during 1951. The huge doors in front and behind the Sabre could be raised in 30 seconds for a scramble. The nose flash, fuselage and tail bands are Yellow with Black stars. (USAF)

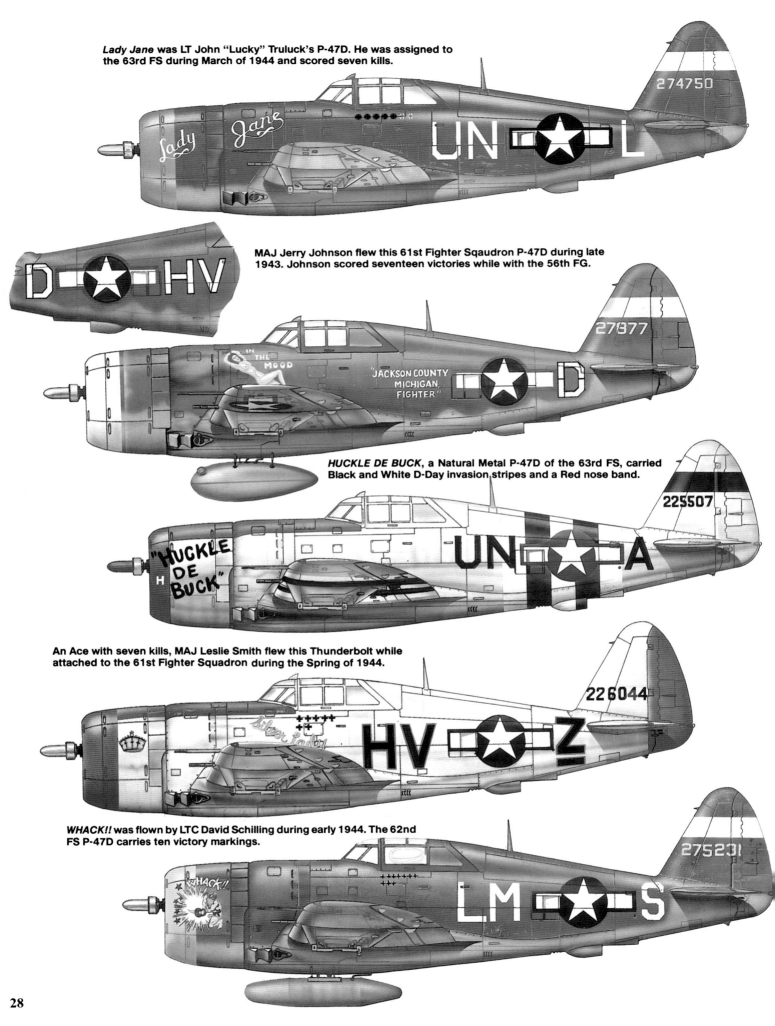

Lady Jane was LT John "Lucky" Truluck's P-47D. He was assigned to the 63rd FS during March of 1944 and scored seven kills.

Lady Jane

274750

UN ★ L

MAJ Jerry Johnson flew this 61st Fighter Sqaudron P-47D during late 1943. Johnson scored seventeen victories while with the 56th FG.

D ★ HV

IN THE MOOD

"JACKSON COUNTY MICHIGAN FIGHTER"

27877

D

HUCKLE DE BUCK, a Natural Metal P-47D of the 63rd FS, carried Black and White D-Day invasion stripes and a Red nose band.

"HUCKLE DE BUCK"

225507

UN ★ A

An Ace with seven kills, MAJ Leslie Smith flew this Thunderbolt while attached to the 61st Fighter Squadron during the Spring of 1944.

226044

HV ★ Z

WHACK!! was flown by LTC David Schilling during early 1944. The 62nd FS P-47D carries ten victory markings.

WHACK!!

275231

LM ★ S

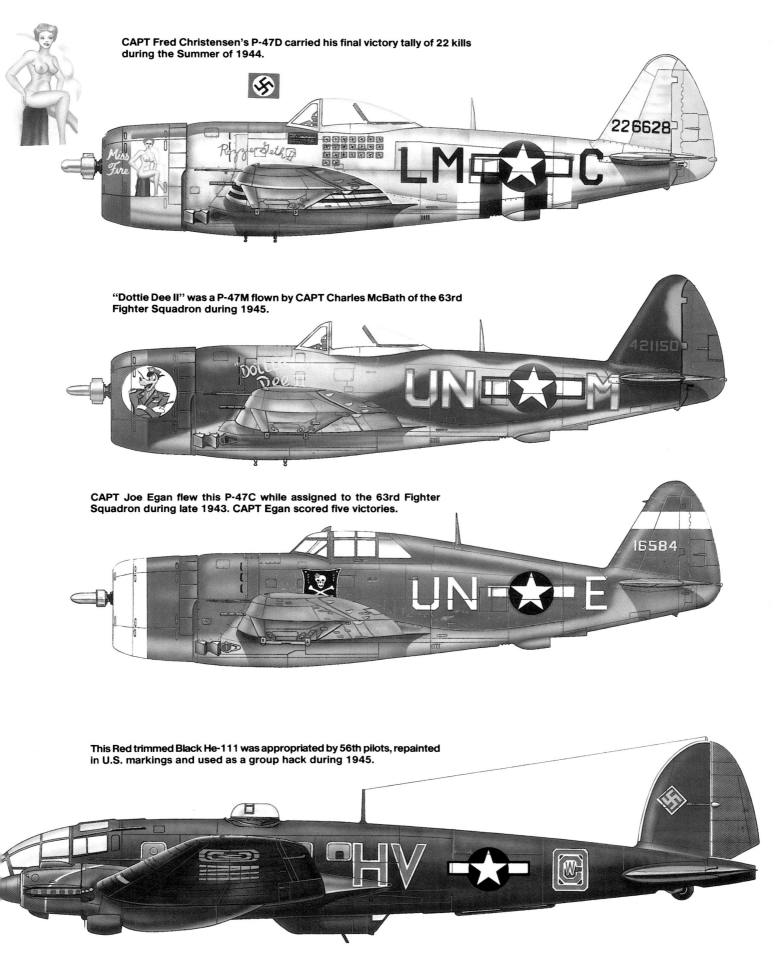

CAPT Fred Christensen's P-47D carried his final victory tally of 22 kills during the Summer of 1944.

"Dottie Dee II" was a P-47M flown by CAPT Charles McBath of the 63rd Fighter Squadron during 1945.

CAPT Joe Egan flew this P-47C while assigned to the 63rd Fighter Squadron during late 1943. CAPT Egan scored five victories.

This Red trimmed Black He-111 was appropriated by 56th pilots, repainted in U.S. markings and used as a group hack during 1945.

Beginning in late 1952, the 63rd FIS, now at Oscoda Air Force Base, began trading in their aging F-86As for F-86Fs. The 63rd FIS was one of a very few Air Defense Command squadrons to be equipped with the F-86F Sabre. The tail marking is a Red and White shooting star. (CAPT J.J. Miller)

The 63rd FIS transitioned to the F-86D during January of 1955, turning in their "day fighter" Sabres for the Sabre Dog. The F-86D was considered the most sophisticated interceptor of the era; however, it was quite a handful for the pilot to handle both the airplane and the intercept equipment. (USAF)

The 62d FIS in F-101Bs had a nuclear capability as they were equipped with the Genie nuclear-tipped "bomber formation destroyer" missile. The Group moved its headquarters to K.I. Sawyer on 1 October 1959, reclaiming the 62nd FIS. But on 25 July 1960, the 61st FIS was again inactivated. On 1 February 1961, the 56th FG (AD) was inactivated and reactivated as a Wing: the 56th FW (AD) (although it had only one squadron, the 62nd FIS with F-101Bs). The Wing made its final move in October of 1963 when the Head-

quarters moved to Duluth, Minnesota. Two months later, on 1 January 1964, the 56th FW (AD) was inactivated again.

They would become an active unit again two years later, but without any of their historic squadrons. When reactivated, the unit would once again go to war. The 62nd FIS remained in service with their F-101Bs at K.I. Sawyer until 30 April 1971, when it too was inactivated. The next mission for the Group would be something very "Special."

As the air defense mission evolved, the 56th FIG responsibility grew and newer technology was developed to handle the enlarged mission. The 61st FIS was equipped with the Lockheed F-94A/B Starfire two seat interceptor. The 61st flew the Starfire until July of 1953 when they converted to the Northrop F-89D Scorpion. (David Menard)

For a short time during 1955, the 42nd FIS was assigned to the 56th FG (AD), flying missions from Chicago. The 42nd FIS and 63rd FIS at Wurtsmith traded bases and aircraft on 18 August 1955 as part of Project *Arrow*. FU-003 was the squadron commander's aircraft and carried Black and Yellow bands around the fuselage. (Marty Isham)

A flight of F-86D Sabre Dogs of the 63rd FIS over Wurtsmith Air Force Base during 1955. The 63rd FIS was not assigned to the 56th FIG at this time, being assigned to the 527th ADG until Project *Arrow* reassigned the squadron to the 56th FIG (AD). (J.J. Miller)

This F-86D of the 62nd FIS was completely opened up for a public display at an Orchard Place AFB Open House during May of 1955. The 62nd FIS flew the F-86D until August of 1959 when they went supersonic with the McDonnell F-101B Voodoo. (David Menard)

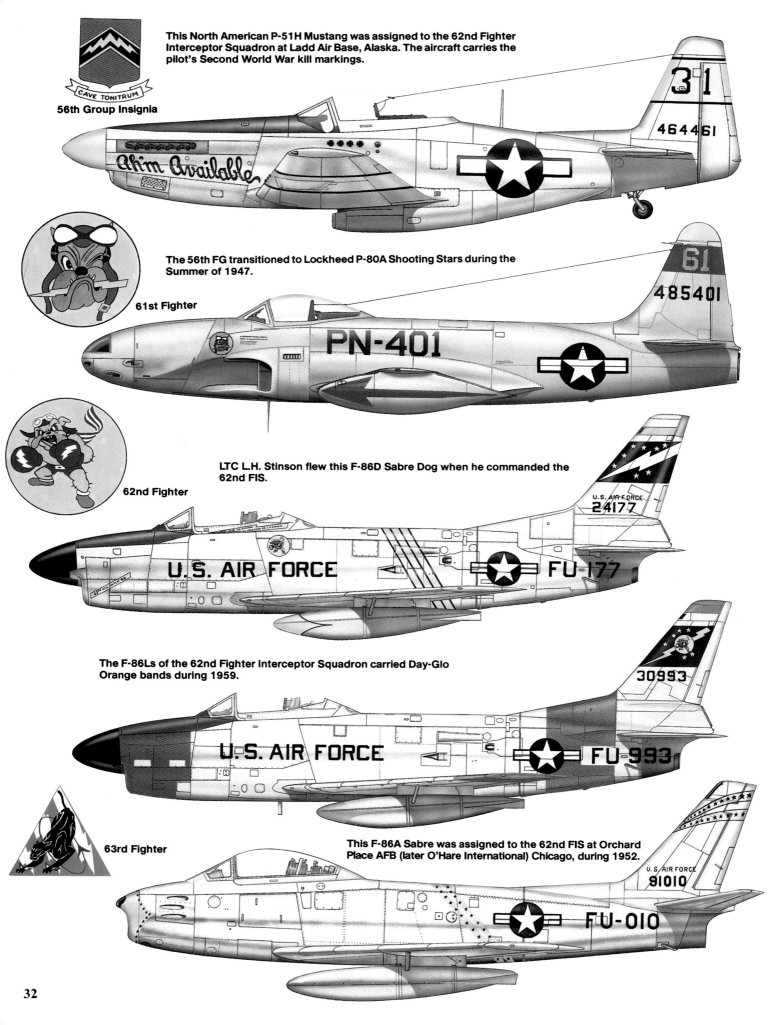

56th Group Insignia

CAVE TONITRUM

This North American P-51H Mustang was assigned to the 62nd Fighter Interceptor Squadron at Ladd Air Base, Alaska. The aircraft carries the pilot's Second World War kill markings.

31
464461
Ah'm Available

The 56th FG transitioned to Lockheed P-80A Shooting Stars during the Summer of 1947.

61st Fighter

61
485401
PN-401

LTC L.H. Stinson flew this F-86D Sabre Dog when he commanded the 62nd FIS.

62nd Fighter

U.S. AIR FORCE
24177
FU-177

The F-86Ls of the 62nd Fighter Interceptor Squadron carried Day-Glo Orange bands during 1959.

U.S. AIR FORCE
30993
FU-993

63rd Fighter

This F-86A Sabre was assigned to the 62nd FIS at Orchard Place AFB (later O'Hare International) Chicago, during 1952.

U.S. AIR FORCE
91010
FU-010

A Lockheed P-80A Shooting Star of the 61st FS on the ramp at Selfridge Field during 1947. (Frank Benneman)

Armorers load 500 pound practice bombs on LTC Ritchie's 61st FS F-80A during the 1949 Gunnery Meet at Las Vegas AFB. (USAF)

An OA-10A Catalina of the 9th Rescue Squadron on the ramp at Selfridge during 1946. The 9th RS was attached to the 56th FG at this time. (Frank Beuneman)

An F-86E of the 61st FIS flys formation off the wingtip of a KB-50 tanker near Detroit during June of 1953. (Ron Picciani)

LTC W.K. Thomas flew this Lockheed F-94B Starfire while assigned to the 61st FIS at Selfridge AFB during 1950. (David Menard)

The 61st FIS transitioned to the Northrop F-89D Scorpion when they were transferred to Ernest Harmon Air Base, Newfoundland under the control of Northeast Air Command during 1955. Some 61st F-89Ds had a sharksmouth similar to that carried on their F-94Bs. (David Menard)

During August of 1959, the 62nd FIS transferred to K.I. Sawyer AFB in upper Michigan, and was assigned to the 473rd FG (AD). Here they transitioned to the F-101B Voodoo. The tail stripes on this overall ADC Gray F-101B are Red and White. (Bill Curry via Marty Isham)

A 62nd FIS F-101B Voodoo on the ramp at K.I. Sawyer Air Force Base during June of 1970. The 62nd FIS was the last of the original 56th FG squadrons still active and was deactivated during April of 1971. (Marty Isham)

Vietnam

During 1967 the Group once again was involved in a shooting war; this time it was halfway around the world in a place called Vietnam. Their mission was vastly different from their previous one of air defense; now the group was flying interdiction missions. The assigned squadrons were all different, and in a modern jet Air Force, the 56th was flying propeller-driven aircraft left over from another era. The 56th Air Commando Wing began flying operations on 8 April 1967 at Nakhon Phanom, Royal Thai Air Force Base, affectionately known as "Naked Fanny" or simply NKP.

At first the 56th ACW was comprised of two squadrons: the 602nd and 606th Air Commando Squadrons (ACS). The aircraft assigned were the Douglas A-1 Skyraider and North American T-28D Trojan counter-insurgency (COIN) aircraft. In September of 1967, another Second World War relic was resurrected and assigned to the 56th when the 609th Air Commando Squadron was assigned to the wing with their Douglas/On Mark A-26A (B-26K) Counter-Invaders. Their mission was to fly interdiction from NKP against the Ho Chi Minh Trail.

Armorers load a 500 pound bomb on an A-1E of the 602th ACS on the ramp at NKP, Thailand. This Skyraider still carries the Navy Gull Gray and White paint scheme. The A-1 was well liked for its endurance and load carrying capability. (AFM)

This A-1E Skyraider was assigned to the 602nd ACS, based at Nakhon Phanom, Thailand. The 56th ACW had detachments throughout Vietnam flying search and rescue missions. This A-1E was *Sandy Lead* of Det 1 at Da Nang. (Tom Hanson via Wayne Mutza)

By the end of USAF operations in Southeast Asia, the 56th ACW (later Special Operations Wing, SOW) would have some ten squadrons assigned, all flying propeller-type aircraft in one form or another. The missions of these squadrons included search and rescue (SAR), interdiction, psywar, reconnaissance, gunship, electronic warfare, and forward air control (FAC). Most of the missions were flown against targets in Laos, Cambodia, and the southern portion of North Vietnam. On 1 August 1968, the 56th ACW was re-designated as the 56th Special Operations Wing (SOW).

The aircraft flown by the 56th SOW were all unique in an all-jet Air Force. The 602d ACS flew various A-1 Skyraider variants. The Skyraider was a Navy attack aircraft used extensively in Korea, and by the Navy from carriers during the early Vietnam years. The 606th ACS flew converted North American T-28 Trojan trainer aircraft. The T-28D-5 (Group II) was a rebuilt and armed version of this high performance propeller driven trainer.

The 609th ACS brought A-26As to South East Asia. The A-26As (B-26Ks) were completely remanufactured prior to the Cuban Crisis of the early 60s. On-Mark Engineering Company in Van Nuys, California contracted with Air Force to remanufacture some forty B-26B and B-26Cs for the COIN role under the designation B-26K. Major

A North American T-28D Trojan of the 606th Special Operations Squadron on the Nakhon Phanom (NKP) ramp during the Fall of 1967. (Norm Crocker)

A 609th Special Operations Squadron Douglas/On Mark B-26K Counter-Invader starts his number two engine on the NKP ramp during 1969. (USAF)

This A-1H of the 602nd Special Operations Squadron stood SAR alert at DaNang during May of 1970. The A-1s would escort rescue helicopters on SAR missions. (LTC Barry Miller)

This 1st SOS A-1E at Nakhon Phanom during 1968 is armed with a PAVE PAT fuel/air explosive bomb. One FAE bomb could clear a 900 foot diameter Landing Zone from the jungle. (AFM)

SOCK IT TO 'EM was a Douglas A-1J Skyraider of the 602nd Special Operations Squadron at NKP, Thailand during 1968. (Don Garrett, Jr.)

The 606 ACS operated this UC-123 flare ship with the call sigh "Lamplighter." The aircraft were used to provide illumiation for B-26s and T-28s operating over the Trail at night. (Cope)

This AC-130 Spectre supported CH-53s of the 21st SOS during the Mayaguez rescue mission. It was joined by 3rd TFS A-7Ds. (JEM)

A 23rd TASS Rockwell OV-10A Bronco on the NKP ramp armed with LAU-3 2.75 inch rocket pods during 1971. The Broncos were used as light strike and FAC aircraft. (Aircraft Publicity Bureau)

This OV-10A Bronco of the 23rd TASS is loaded with 2.75 inch rocket pods. Normally, these pods would carry smoke rockets used in marking targets for tactical fighters. (56th TTW/HQ)

modifications included a rebuilt fuselage, tail and wing assembly, along with a larger rudder. The wings were reinforced and had eight underwing weapons pylons and permanently mounted wingtip fuel tanks. The six .50 caliber machine guns originally carried in the wings on the B-26B and B-26Cs were deleted to save weight. The upper and lower remote controlled twin machine gun turrets were also deleted.

The B-26K was powered by two 2,500 hp Pratt & Whitney R-2800-103W radial engines housed in modified cowlings, driving three blade reversible propellers. Internally, the A-26As had dual flight controls, heavy duty anti-skid brakes, a quick change nose carrying eight .50 caliber machine guns, greater capacity DC generators and inverters for the additional electric power needed to supply the many electronic and radio systems.

The underwing pylons could carry a wide variety of weapons including GE SUU-11 Minigun pods, rocket launchers, CBUs, napalm, bombs and 230 gallon drop tanks. The internal bomb bay could carry either 4,000 pounds of bombs, electronic and reconnaissance gear, or a 675 gallon ferry tank. The gross weight rose from 36,000 to 43,300 pounds, top speed went from 314 to 345 mph, rate of climb rose from 2,745 ft/min to 2,990, service ceiling went from 23,000 to 30,000 feet and combat range rose from 210 to 500 miles (800 miles with drop tanks). The A-26A crews would need all these capabilities

A *Zorro* T-28D flies over the Laotian jungle during 1967. The T-28D had strengthened wings to handle six underwing weapons pylons and .50 caliber machine gun pods. The aircraft did not carry national insignia and were camouflaged in two tone Green and Gray uppersurfaces. (Norm Crocker)

The 606th ACS flew the North American T-28D Trojan trainer modified for the counter-insurgency (COIN) mission. The *Zorro* T-28Ds primarily flew interdiction missions against truck traffic on the Ho Chi Minh Trail from 1967 through June of 1968. (USAF)

to survive their mission.

Their call sign was *Nimrod* and their mission was night interdiction on the Ho Chi Minh Trail, that is truck-busting at night! The Ho Chi Minh Trail ran from the Mu Gia Pass in North Vietnam to the Mekong Delta, passing through Laos and Cambodia. The Trail ran up and down the 5,000 foot mountains (known as karsts), through river valleys and deltas. Additionally, most of it was covered with dense jungle. The North Vietnamese (NVA) moved most of their supplies and personnel into South Vietnam via The Trail and most of that movement was at night.

It was the job of the *Nimrod* A-26As to stop it. The NVA knew this and defended The Trail with multiple batteries of radar-directed anti-aircraft weapons. The guns ranged in size from 12.7MM (.50 caliber), up through the 57MM anti-aircraft artillery. Later, when AC-130 and AC-119 gunships starting roaming the Trail at night and began making life tough for NVA truck traffic, Soviet SA-2 surface-to-air missiles began appearing along The Trail. The Trail was the NVA "lifeline" for their troops in South Vietnam and had to be kept open.

A pair of *Zorro* T-28s patrol over Laos during late 1967. One of the modifications to the T-28D-5 was the whip antenna under the fuselage. This was the antenna for an FM radio used to communicate with ground troops. (Norm Crocker)

A *Zorro* pilot preflights his T-28D-5 on the NKP ramp. The aircraft is carrying an SUU-14 bomblet dispenser on the outboard wing pylon and a Mk 82 low drag bomb on the inboard pylon. (Norm Crocker)

A *Nimrod* mission was never dull or routine. You were assigned a specific stretch of road and a FAC to work with. The rules were simple, you just flew up and down that stretch of road looking for something to shoot at. Sounds simple if you're interdicting the San Diego Freeway — nice and flat, wide-open and all the vehicles have their lights on. But The Trail was vastly different. The *Nimrod* flew slightly high, so as to clear all the natural obstacles (trees, ridge lines, etc.). The FAC aircraft, O-2s, UC-123 and HC-130s, would use various night observation devices to locate the elusive enemy. Starlight scopes, infrared seekers, low-light-level TV were all used along The Trail.

As soon as a FAC located a possible target, he would mark it for the A-26 with smoke rockets, flares or napalm. The *Nimrod* crew would then decide what ordnance to use: bombs, rockets, CBUs, napalm, etc. After each pass the FAC would take another look and assess the damage. If a large target such as a full truck park or convoy was located, additional A-26s from other portions of The Trail were called in, or night flying A-1s and T-28s. They would bomb the target

area until the FAC called the target destroyed, or all their weapons were used up.

A typical *Nimrod* mission started before dusk. You preflighted the aircraft for about an hour, making sure that all the ordnance was properly wired and all the electronics worked. The last was the most important, since it could easily save your life. Around dusk, the *Nimrod* would take off and rendezvous with his FAC, usually an O-2. Then you began trolling up and down the section of The Trail that was assigned to you. More often than not, you searched your section without sighting a single thing. The NVA only moved their trucks in short spurts, always on the watch for aircraft.

Sometimes the FAC would not arrive on station, or would have to turn back. At that point the *Nimrod* crew would simply use their eyes to locate targets. Sometimes it was easy, such as when the NVA would get careless and leave a camp fire burning. Or when a large caliber AAA battery would open up and you could note the flash. With AAA you played it very cautious since the A-26A was at a distinct disadvantage with radar-directed AAA. The *Nimrod* would usually come in low and fast, then string out some CBUs triangulating the target area. If you were lucky maybe you would hit an ammo

The first *Nimrod* B-26Ks taxi to their parking spots at NKP during September of 1967. The B-26K was a completely rebuilt and refurbished Douglas B-26B with strengthened wings mounting eight weapons pylons, new engines, a taller rudder and modern electronics. (USAF)

The major force behind the *Zorro* successes was the personnel that kept them flying. MAJ John Pattee (4th from left kneeling) was the *Zorro* section commander. To his left is CAPT Norm Crocker, who contributed the *Zorro* material for this book. The flight crews wore Black flying suits with a *Zorro* mask and emblem on the suit. (Norm Crocker)

The 56th TFW converted to the F-4D Phantom during October of 1977. This Phantom was flown by COL Henry Canterbury during 1980 when he was commander of the 56th TFW. (Marty Isham)

F-16 Fighting Falcons of the 61st TFTS on the ramp at McDill Air Force Base during 1986. The 56th is now strictly a training unit, training air and ground crews for Tactical Air Command units. (56th TTW/HQ)

This F-4D of the 62nd TFS on the ramp at Nellis Air Force Base took part in the 1979 Red Flag exercises. Red Flag is the most realistic set of exercises ever conducted by the USAF. (Mick Roth)

COL Jimmy Cash, commander of the 56th TTW, flew this F-16A during 1985. The 56th TFW became a Tactical Training Wing on 1 October 1981. (JEM Aviation Slides)

The *Nimrod* B-26Ks began operations from NKP during September of 1967. *Nimrod* was the call sign for the 609th Air Commando Squadron's B-26s. Flying Second World War/Korea veteran aircraft types, the 56th ACW soon became known as the "Antique Air Force." (USAF)

dump or even the gun itself. If not, and the gun didn't get you, you would call in other FACs and *Nimrod* to continue the attack until everyone was satisfied.

The *Nimrod* rarely bombed the road net itself. That was useless since The Trail had numerous bypasses and hidden roads. The best mission was when you were lucky enough to locate a supply dump or truck "park." A *Nimrod* crew would really feel good if they could leave six or eight trucks burning. Charlie would hurt from that mission.

The 606th Air Commando Squadron was a composite squadron with a variety of aircraft types and missions. They operated UC-123 aircraft, call sign *Lamplighter*, in the night FAC role. These aircraft also flew flare missions in support of the Cambodian and/or Laotian armies. U-10 aircraft, call sign *Litterbug*, flew psywar missions dropping leaflets and flying administrative missions. But it was the North American T-28D-5 Trojans, call sign *Zorro*, that flew the bulk of the 606th missions.

The *Zorros* flew both close air support for the Laotian Army and Trail interdiction missions. The T-28D-5 aircraft were modified with two underwing .50 caliber machine gun gondolas fed from ammunition bins within the wing. The aircraft also had a total of six weapons pylons capable of carrying a wide variety of ordnance including gun and rocket pods, bombs, CBUs, napalm, and flares. A number of the T-28s were also equipped with the Yankee rocket extraction seat. This escape system allowed the pilot to quickly leave a damaged aircraft, even at low altitudes and low speeds. Additionally, the T-28s were equipped with FM radios for communicating with ground troops, with a long whip antenna being mounted on the fuselage underside.

The main *Zorro* mission was Trail interdiction day and night. They also flew daylight sorties such as escort missions for helicopters inserting intelligence teams throughout South East Asia, including North Vietnam. Another mission that fell to the *Zorros* was pilot training of Cambodian and Laotian aircrews on the T-28. Detachment 1 at Udorn, Thailand, call sign *Tiger* had the training responsibility and many of the T-28s flew in Lao insignia, while still others carried no markings except the aircraft serial number. The Americans were only supposed to train friendly aircrews in the operation of the T-28, but they often ended up flying combat missions.

A *Nimrod* B-26K on the PSP ramp at NKP, Thailand during late 1967. The aircraft is carrying a multiple ejector rack on the wing pylon loaded with parachute flares. The 609th ACS tail code letters, TA, had not been painted on the fin at this time. (Robert F Dorr)

One ex-606th Master Sergeant recalled that a T-28D-5 of Det 1 at Udorn, Thailand scored a MiG-17 kill over Laos, north of Na Khang. The MiG-17 tried to turn low and slow with the T-28 and got in front of the Trojan's twin .50 caliber machine guns. The kill was witnessed by several F-4 Phantoms of the 13th Tactical Fighter Squadron that were at a higher altitude just waiting for the MiG-17 to make a break for North Vietnam so they could engage it. When the *Zorro* pilot returned to base, he did a victory roll over the F-4 parking area.

The T-28D-5 *Zorro* was an excellent aircraft for the COIN mission.

MAJ Norm Crocker recalled:

> It was better than the A-1E or A-1G (two seaters) and worse than the A-1H or A-1J (single seaters). For the night mission, it was

A T-28D-5 Trojan of the 606th Air Command Squadron parked on the NKP ramp. The T-28s had problems with the wings. The aircraft were not designed for combat and over time the wings would become over-stress from the weight being carried and sometimes failed in flight. (Norm Crocker)

A *Zorro* T-28 taxis to the active runway at NKP armed with an SUU-14A, a Mk 82 bomb and a LAU-32 rocket pod, for a daylight mission in support of Laotian Army units during the Battle of Thakhet. The armament specialist has just removed the safety wires form the weapons. (Norm Crocker)

> essential to be able to see how the NVA guns were tracking you. You could not do that with the two seat Skyraiders but you could in the T-28. The T-28, however, had two major problems: engine failures and weak wings. The engine failures in SEA were quite bad during early 1967. A failed engine at night over The Trail could cause some rather exciting moments. You would really start talking to yourself at times and to the crew chief when, and if, you got back. That problem was finally solved, but not until after the T-28s were phased out of USAF service in favor of the A-1s.
>
> The second problem was much harder to control. Even with the rebuilt and strengthened wing, they were still weak for a combat mission (the T-28 was never designed for the stress of combat).

Det 1 of the 56th SOW at Udorn, Thailand, was responsible for the flight training of Laotian pilots in the T-28D fighter-bomber. *Zorro* pilots performed this task under the call sign *Tiger*, often flying combat missions in the Laotian marked aircraft. (Norm Crocker)

You weren't supposed to pull over 4 Gs in a T-28. But when the guns are laying all over you, the rules go out the window. Between 1967 and 1973, several T-28s were lost due to violent combat maneuvers that resulted in the wings coming off the aircraft. It was hard to prevent, but we did get it under control during 1973 by replacing the aluminum spar cap ends with steel ones. Would I fly a Zorro T-28 again? A resounding YES!

How good were the *Zorros*? A *Nimrod* pilot once remarked "If a *Zorro* can find a truck, he'll get it every time!" The *Zorros* were just as deadly in the daylight. On 11 March 1968, the T-28s flew daylight support during the Battle of Thakhet and were credited as the major force in halting the advance of hostile troops.

On 9 June 1968, MAJs Roland Vernon and Price Harris flew the final USAF T-28 *Zorro* missions. The *Zorros* then transitioned into the A-1 Skyraider. In October of 1968, the 606th ACS was split into two squadrons - the 606th and 22nd Air Commando Squadrons, both flying Douglas A-1 Skyraiders. The 1st Air Commando Squadron, also flying A-1s was added to the 56th during December of 1967.

Aircraft and mission types continued to expand throughout the war.

The 602nd ACS, call sign *Sandy* or *Firefly*, flew at least four different variants of the Douglas A-1 Skyraider in two separate and very distinct missions. *Firefly* missions were twenty-four hour alert missions flown in support of the Royal Laotian Army. The *Fireflies* usually flew with Cessna O-1 and O-2 Forward Air Controllers (FACs), pounding the Pathet Lao positions with Cluster Bomb Units (CBUs), napalm or Minigun pods. The Minigun pods were especially effective against Pathet Loa troops attacking in the open. At night, the A-1s would also fly interdiction missions along the Trail.

But it was the *Sandys* that gained notoriety both in the press and throughout the Air Forces operating in South East Asia. The *Sandys* were the guys that flew Search and Rescue (SAR) escort missions anywhere in SEA. Whenever an aircraft went down, a massive effort was put on to rescue the downed crew. It involved great teamwork, especially if the pilot went down inside North Vietnam. The NVA wanted that pilot as badly as we did and their efforts to capture the pilots were almost as great as those of our rescue forces to extract them.

SAR aircraft teams stood twenty-four hour alerts in revetments on bases scattered throughout South East Asia. The SAR team usually consisted of a pair of Sikorsky HH-3 Jolly Green Giant

Det 1 and Lao T-28Ds undergo maintenance in a hangar at Udorn. The overall Light Gray Lao T-28 carries a Black pony marking on the fin, while the camouflaged T-28 carries no markings except its serial number on the fin in Black. (Norm Crocker)

Camouflaged and Light Gray *Tiger* T-28s share the ramp at Udorn, Thailand. The USAF provided Laos with number of T-28Ds under the Military Assistance Program. The aircraft were normally flown by American pilots with Lao observers. (Norm Crocker)

A portion of the *Nimrod* detachment poses in front of one of their B-26K Invaders at NKP. The B-26 carries at least sixty-five truck kill markings painted on the fuselage side below the nose. The eight nose guns are taped over to keep out dust and moisture. (AFM)

helicopters and at least one flight of four A-1 *Sandys* to serve as escorts (later in the war, the HH-3s were replaced with HH-53 Super Jollys and the A-1s were replaced by A-7Es). The SAR team sometimes had a FAC O-2 or OV-10 assigned to serve as its eyes, but usually not deep into North Vietnam. As soon as a pilot called out that he had been hit, the SAR team would be airborne and heading in his direction. It was the job of the *Sandy* Skyraiders to suppress any type of enemy action that might interfere with the Jolly Green helicopter being able to get in, extract the downed airman, and get back out.

The normal mission profile would have the *Sandy* flight split when nearing the location of the down crew. *Sandy 3* and *Sandy 4* would remain with the Jolly Green, while the *Sandy 1* and his wingman would proceed to the target area first. The mission of *Sandy 1*

would be to find the downed crew and locate enemy defenses. The A-1 would fly over the area and advise the downed crew to call when he passed overhead, the Skyraider would then continue ahead for several minutes before turning (to keep the VC/NVA from getting a fix on the crew location). After fixing the location of the downed crew, *Sandy 1* would begin searching for enemy guns.

This was perhaps the most dangerous part of the entire mission. The Skyraider would fly pass after pass over the area, trying to get the VC/NVA to fire on him. If they did, then the FAC (or his wingman) would fix their locations and *Sandy 1* and *Sandy 2* would attack them. The Skyraiders could carry a variety of weapons in addition to their internal armament of four 20MM cannons making them ideal for this mission.

Once the defense were silenced or softened, the HH-3 Jolly Green would move in for the actual rescue. While the helicopter

Batgirl appeared on the nose of a 609th Special Operations Squadron B-26K Invader (64-17645) at NKP during the Summer of 1968. The figure was based loosely on a character in the popular TV series "Batman" which was shown on armed forces TV. (AFM)

CAPTs Tengan and Nelson from the *Nimrod* detachment pose in front of a 609th B-26K at NKP during 1968. The aircraft, TA-677, has truck kill markings painted on the entire fuselage from the nose to the trailing edge of the wing. There are at least 99 visible. (AFM)

A 609th SOS B-26K on patrol over Laos during the Summer of 1968. The B-26 is armed with napalm bombs, SUU-14A dispensers and flares on the outboard MER. The B-26K Counter Invaders retained the eight .50 caliber machine gun nose of the B-26B but deleted the six wing guns. (AFM)

made the pickup, the A-1s would continue to orbit the area and watch for enemy movement. If large formations of the enemy were spotted, than *Sandy 1* might call in some *Fast Movers* for assistance. In South Vietnam these were usually F-100s Super Sabres (early in the war) while for missions over the north, F-105s Thunderchiefs or F-4 Phantoms were more usual.

Support aircraft for the *Sandy* mission were many and varied, depending on where the pilot went down. A SAR mission in South Vietnam, Laos, or Cambodia might call for a SAR team, a FAC, maybe an EC-47 to monitor the VC radio traffic, and perhaps some Army UH-1 Huey gunships to help the *Sandys* soften up the defenses. Perhaps an AC-47 Spooky might be called in if the area became too "intense."

An "out-country" (North Vietnam) mission was different. Instead of a FAC O-2, the eyes of the SAR team would more likely be either an OV-10 or a "fast FAC" two seat tactical jet fighter. If the threat allowed, the team might be assisted by an AC-130 Spectre gunship, with its sophisticated sensors. A major plus was the additional and very awesome firepower that was available with the AC-130.

If a mission went further into North Vietnam, with its highly sophisticated air defense system, specialized mission aircraft were added to the SAR force. EC-121 Warning Stars would monitor any North Vietnamese Air Force MiG fighter traffic operating near the SAR effort. Douglas EB-66s would be used to jam NVA radars tracking the SAR team. Republic F-105F Wild Weasels would suppress or knock out the SAMs and AAA around the downed pilot, while Lockheed HC-130 Hercules tankers would refuel the Jolly Greens. Any aircraft from the strike force with remaining ordnance and fuel would orbit near the downed pilot, waiting to be used at the discretion of the SAR mission commander. But it was still the *Sandys* and Jolly Greens that went in "low and slow" to actually retrieve the downed airman.

One of the more memorable Skyraider missions involved an A-1E of the 602nd and another Skyraider of the 1st Air Commando Squadron. MAJ D.W. Myers of the 602nd was hit while providing close air support for a Special Forces camp in the A Shau Valley and crash landed his A-1E on the camp's runway. Circling above, MAJ Bernard G. Fisher realized that Myers would probably be taken prisoner if something was not done to rescue him quickly. Organizing the remaining Skyraiders to supply support for the rescue, Fisher landed his Skyraider on the debris covered runway. Myers quickly boarded Fisher's A-1 and they took off again under heavy VC fire.

This 56th Special Operations Wing A-1E Skyraider is armed with a Minigun pod on the inboard pylon, and seven shot rocket pods on the outboard pylons. Along with its internal four 20MM cannons, the Minigun pod gave the Skyraider devastating firepower. (Tom Hanson via Wayne Mutza)

The typical armament load of a 1st ACS A-1 flying the search and rescue (SAR) mission was (from left) a Cluster Bomb Unit, a LAU-32A rocket launcher, a 250 pound bomb, and two SUU-14A bomblet dispensors. The port wing load will be the same except that an SUU-11 Minigun pod will be hung on the port wing tank pylon. (Tom Hanson via Wayne Mutza)

An A-1G Skyraider of the 602nd Special Operations Squadron. Air Force Skyraiders were camouflaged in Tan and two tone Green upper-surfaces over Light Gray undersurfaces in what became known as the Southeast Asia camouflage scheme. The size of the national insignia was also reduced on camouflaged aircraft.

For this action MAJ Fisher was awarded the Medal of Honor, the first to be won by an airman in Vietnam.

Another 602nd SOS Skyraider pilot, LTC William A. Jones, was also a recipient of the nation's highest award for a *Sandy* mission over North Vietnam. LTC Jones was flying as *Sandy Lead* on 1 September 1968 on a rescue mission to pick up the crew of a downed F-4. Enroute to the scene, he learned that one of the Phantom crew had been captured and the other was trying to evade the NVA. While directing the rescue force, Jones' Skyraider was hit by ground fire. Even though his A-1 was damaged, Jones pressed home his attack on the NVA guns near the downed Phantom crewman, knocking out one site with 20MM cannon fire and another with Cluster Bomb Units. During this attack, his A-1 was hit again.

The anti-aircraft fire had ignited the rocket motor of the Yankee escape seat, starting a fire in the cockpit and burning Jones' legs. He blew the canopy intending to bail out; however, the damaged seat would not function. His Skyraider was now burning out of control and his radio died just as a relief *Sandy* flight arrived on scene. Jones returned to NKP, made a straight in landing and was rushed to an ambulance, badly burned. While being loaded into the ambulance, Jones continued to pass on information to be relayed to the rescue forces, information that led to the successful rescue of the second Phantom crewman. For his efforts, "above and beyond the call of duty," LTC Jones was awarded the Medal of Honor.

The Skyraider continued to be the leading aircraft of the 56th and was well liked, although by 1969, its age was beginning to show. Pilots reported that the wing cannons were developing a habit of jamming and exploding and that the pneumatic tail wheel caused problems when operating from the pierced steel planking at NKP.

A-1s out of NKP covered the Son Tay rescue mission during November of 1970, but by 1972 the three units operating out of NKP had been reduced to one, the 1st SOS. The expansion of the Vietnamese Air Force led to a draw down of A-1s in USAF units. To free up A-1s to be passed to the VNAF, the *Sandy* units began to re-equip with the LTV A-7D Corsair II. The final A-1 *Sandy* mission was flown on 7 November 1972.

This camouflaged A-1E Skyraider of the 602nd Special Operations Squadron warming up its engine on the ramp at Nakhon Phanom (NKP), Thailand, has the tail code and aircraft serial number in White. (Tom Hanson via Wayne Mutza)

This A-1E Skyraider of the 602nd SOS is parked on the open ramp ahead of the blast revetments at NKP. The 602nd deployed detachments of A-1s to various locations throughout the war zone to stand SAR alert. (Tom Hanson via Wayne Mutza)

CAPT Norm Crocker returns to NKP after checking out one of the "new" A-1Gs during 1968. The night operations air crews liked the two-seat A-1E/G over the single A-1H/J as it offered a second pair of eyes to scan for AAA fire along the Trail. (Norm Crocker)

1st ACS A-1Es on the ramp at NKP, Thailand. Capable of carrying every conceivable type of weapon from Gatling guns to fuel-air explosive (FAE) bombs, the A-1s were the perfect aircraft for search and rescue escort because of their long loiter time and huge load carrying capacity. (Norm Crocker)

BLOOD, SWEAT, TEARS was an A-1H of the 602nd Air Commando Squadron. The aircraft had the Black squadron code letters and aircraft serial number outlined in White. The Skyraider is armed with rocket pods, bomblet dispensers, 250 pound bombs and a Minigun pod. (USAF)

An A-1H of the 602nd SOS shares the ramp at NKP with another Skyraider variant used by the squadron, a two seat A-1E. The squadron code letters, TT, and serial number are carried on the fin in Black. (Cope)

A 602nd SOS A-1H Skyraider sits SAR alert at DaNang during 1970. The SAR A-1s were not the most pristine aircraft in appearance, but the maintenance of mechanical parts was excellent at all times. It was said that an A-1 pilot would be worried if the underfuselage tank was not covered with oil. (MAJ Barry Miller)

A heavily armed *Sandy* A-1 Skyraider is parked in a blast revetment on an air base in Southeast Asia. SAR detachments of A-1s were stationed at various locations throughout SEA including Laos, South Vietnam and Cambodia. This Skyraider was named *MISS DOREEN* and was flown by the Squadron Commander. (USAF)

Miss Kate and *White Rabbit*, a pair of A-1Hs of the 1st SOS share the ramp at NKP, Thailand, during late 1967. The aircraft are both armed with unfinned napalm bombs which had to be dropped from very low levels to be accurate. (Tom Hanson via Wayne Mutza)

An A-1H Skyraider of the 1st Special Operations Squadron (1st SOS) on patrol over Laos. The aircraft is carrying SUU-14A bomblet dispensers on the outboard pylons and a Minigun pod on the inboard pylon. (JEM)

Expansion

The 56th, which became a Special Operations Wing (SOW) on 1 August 1968, had some ten squadrons attached to it by the time the war "ended." These included the 18th Special Operations Squadron with Fairchild AC-119K Stinger gunships, the 21st Special Operations Squadron flying Sikorsky HH-3 Jolly Greens, the 23rd Tactical Air Support Squadron in Cessna O-2s and Rockwell OV-10 Bronco Forward Air Controller aircraft, the 361st Tactical Electronic Warfare Squadron in Douglas EC-47s and the 554th Reconnaissance Squadron flying Lockheed EC-121 Warning Stars. Each squadron had detachments scattered throughout South East Asia.

The 18th SOS flew the Fairchild AC-119K (call sign *Stinger*) gunship for operations over the Ho Chi Minh Trail. The AC-119K was armed with four 7.62MM Miniguns and two 20MM Vulcan cannons and had a number of night vision and other sensors. The mission of the AC-119s was truck busting on the trail and fire support for outlying bases in the northern portion of South Vietnam. The AC-119K was an improvement over earlier AC-119 gunships both in firepower and performance. With two J85 turbojets mounted in underwing pods, the AC-119K could lift more weight and was faster.

The crew of the AC-119K consisted of ten men: aircraft commander, co-pilot, navigator/safety officer, FLIR/radar operator, NOD operator, flight engineer, three gunners, and an illuminator operator. The first AC-119Ks arrived in Vietnam on 3 November 1969, assigned to the 14th SOW and by the end of the year, the unit's full inventory of twelve aircraft were on hand and ready for operations. To better utilize the aircraft, the unit was split into flights and deployed to three different bases in South Vietnam. Flight A with six aircraft was deployed to Da Nang Air Base, Flight B with three aircraft was sent to Phu Cat, while Flight C remained at the main support base of Phan Rang. Later, the squadron was further split with three aircraft deploying to Udorn, Thailand as Flight D.

By February of 1970, the 18th SOS was flying some ten sorties per day. These included armed reconnaissance, close air support and truck busting. The AC-119Ks flew missions over Cambodia, hunting truck and sampan traffic. In nine months of operations over Cambodia, the AC-119s accounted for some 600 vehicles. Operations in Laos saw Flight D AC-119Ks, operating out of Udorn, account for some seventy percent of the trucks destroyed during February and March of 1970.

The 21st SOS flew Sikorsky HH-3E *Jolly Green Giant* helicopters deep into North Vietnam to rescue downed air crews. A Search And Rescue (SAR) team usually consisted of a pair of HH-3s with at least one flight of four A-1 *Sandys*. (AFM)

There were several other units attached to the 56th SOW at NKP, including reconnaissance and gunships squadrons. This AC-119K *Stinger* of the 18th SOS flew Trail interdiction missions during the early 1970s. In 1972 the 56th SOW had seven squadrons under its control. (J. Ward Boyce)

During October of 1970, Flight D moved from Udorn to Nakhon Phanom, Thailand with six AC-119Ks. The *Stingers* were highly effective and by the time the war ended, the 18th SOS had a confirmed kill tally of some 2,206 trucks.

The 21st SOS flew the Sikorsky HH-3, call sign *Jolly Green*, helicopter for search and rescue ,missions, working with the *Sandys*. The HH-3E was the first rescue helicopter to be air refuelable. Additionally, the HH-3s carried M60 machine guns and Miniguns for self defense during the actual pickup. Later in the war the 21st converted from the HH-3 to the HH-53 Super Jolly Green. The HH-53Cs were bigger, faster, longer ranged, air refuelable, and more heavily armed. Each HH-53 carried an internal armament of three Miniguns, two in the forward fuselage doors and one firing from the rear ramp. These guns provide the helicopter with good suppressive firepower.

During May of 1975, the 21st SOS found itself involved in the rescue attempt for the crew of the merchant ship, MAYAGUEZ. The ship was seized by Cambodian forces and the crew was taken ashore at Koah Sam Leom island. It was decided to rescue the crew and a force was quickly assembled to carry out this mission. The Super Jolly Greens would be used to transport Marines form Thailand out to the island, under the call sign *Knife*. The first two helicopters to touch down on the western beach of the island, *Knife 21 and Knife 22*, met

A CH-53 *Super Jolly* on the ramp at NKP during 1970. The CH-53s had the call sign *Knife* and were armed with 7.62MM Miniguns on flexible mounts in the forward entry doors and on the rear ramp. This CH-53 is not equipped with a refueling probe. (56th TTW/HO)

with heavy fire. *Knife 21* was heavily damaged and crashed after making a single engine liftoff from the island. The survivors were rescued by another CH-53. Two other helicopters were also seriously damaged.

On the eastern beach, the situation was almost as bad. Two helicopters, *Knife 23* and *Knife 31*, were both shot down with their crews and the Marines they carried being pinned down on the beach by heavy fire from the Cambodian defenders.

After the Cambodians surrendered the crew of the MAYA-GUEZ, the decision was made to extract the Marines from the island and terminate the operation. In order to extract the Marines from the eastern beach, additional CH-53 and HH-53 missions would have to be flown. Throughout the operation, OV-10s from another 56th SOW unit, the 23rd TAAS, provided support and FAC services, marking targets for air strikes. Under the cover of tactical air support from USAF A-7Ds, AC-130s and F-4Es, backed up by Navy F-4Ns and A-7Es, three H-53s (two HH-53s and one CH-53) successfully recovered the Marines. The last helicopter off the island was *Knife 51* which extracted the last twenty-nine men from the western beach.

The 554th Reconnaissance Squadron operated the several variants of the EC-121 Warning Star, including the EC-121R. The EC-121R differed from other Warning Star variants in that it did not carry the large radomes. It was mainly used as a communications relay aircraft which relayed transmissions from *Igloo White* sensors planted on the Ho Chi Minh trail back to the main control center at NKP. These aircraft were camouflaged and operated at low level. Later the EC-121Rs were replaced by QU-22B drones because their large size had made them vulnerable to ground fire.

The 361st Tactical Electronic Warfare Squadron flew a variety of EC-47 variants including the EC-47N, EC-47P and EC-47Q. These aircraft filled a variety of roles including communications intercept (Project *Hawk Eye*), signals intelligence, jammers, and communications relay. The Project *Hawk Eye* aircraft were involved in direction finding missions aimed at pinpointing the locations of VC and NVA radio transmitters. Once located and identified, the radio could often be tied to a VC/NVA unit, providing the field commander with invaluable intelligence on the enemy's disposition.

The mission of the 23rd Tactical Air Support Squadron was to provide Forward Air Controller services to the tactical fighters. The FACs would patrol an area searching for enemy activity. The squadron began operations flying the Cessna O-2A Super Skymaster and later the Rockwell OV-10A Bronco. A typical O-2 mission would last some five hours and was flown at very low level. FACs operated both

The HH-53 was equipped with a refueling probe mounted on the starboard side and sponson mounted external fuel tanks. HH-53s operated alongside *Knife* CH-53s during the MAYAGUEZ incident. (USAF)

An HH-53 *Super Jolly Green Giant* refuels from an HC-130H tanker over the South China Sea. The HH-53 gave the SAR teams a far better mission range, speed and payload than the older HH-3E. Besides serving as the tanker for the HH-53, the HC-130H would also act as an airborne command post for the SAR mission. (AFM)

This weathered overall Gunship Gray AC-130 Spectre of the 16th SOS operated out of Korat, Thailand during the MAYAGUEZ rescue mission to support the *Knife* CH-53s and *Jolly* HH-53s. The gunships sank several Khmer gunboats. (Don Jay)

in South and North Vietnam. Missions into North Vietnam normally were confined to the southern portions of the country.

For missions into North Vietnam, the O-2A would carry a full load of fuel, two Minigun pods on the inboard pylons and two rocket pods on the outboard pylon. The rocket pods held WP (white phosphorus) rockets for marking targets. In North Vietnam the FAC would search for targets such as truck convoys, SAM sites, troop and supply concentrations, and other worthwhile targets. The O-2s flew from 5,000 to 7,000 feet and at speeds up to 115 mph. Once a target is found, the FAC marks it with a WP rocket and calls in the fighters to take it out.

Besides the missions over South and North Vietnam, the 23rd TASS had a number of O-2As that were painted in an overall Black camouflage for night missions over the Ho Chi Minh Trail. During the course of O-2 operations in South East Asia, eighty-two aircraft were lost in combat and another twenty-two were lost in accidents. Of these losses, at least three were shot down by SA-7 shoulder launched surface-to-air missiles.

The 23rd also operated the Rockwell OV-10A Bronco, an aircraft specifically designed for the counterinsurgency mission. The Bronco was faster and better armed than the O-2s although it has slightly less endurance. The OV-10A had an internal armament of four 7.62MM M60C machine guns and seven weapons stations (one centerline, two underwing and four on the sponson) capable of carrying a total of 3,600 pounds of ordnance. For night FAC missions, the OV-10As of the 23rd were modified under the *Pave Nail* program. The modification consisted of the addition of a laser rangefinder and target designator mounted in a pod under the fuselage and a stabilized night vision sight. Reportedly, some fifteen aircraft were modified to the *Pave Nail* configuration (they were all later reconverted back to standard OV-10A configuration during 1974).

Pave Nail operations consisted of two OV-10s operating as a team. One aircraft would operate at high level, acting as a manager. This Bronco would brief incoming tactical aircraft and direct them to the other OV-10 which was flying at a lower level marking targets with the laser.

The Broncos remained in operation until June of 1974 when the last twelve aircraft were ferried out of NKP to Osan Air Base, Korea, where they were turned over to the 19th TASS.

On 30 June 1975, the 56th SOW closed operations at NKP and the "antique air force" was phased out.

The EC-121 Warning Star served in a variety of configurations and missions including airborne early warning, traffic control and electronic intelligence. Besides the Air Force, EC-121s were flown in South East Asia by the Navy and Air National Guard units. (USAF)

Another unit at NKP that worked the night missions with the 56th ACW A-1s, T-28s and B-26Ks, was the Royal Laotian Air Force AC-47 squadron. Laotian AC-47s differed from USAF and VNAF gunships in their armament. These AC-47 carried three side-firing .50 caliber machine guns. (Norm Crocker)

If an *Igloo White* sensor picked up a scent or noise, it would send out a radio signal. A QU-22 from the 554th Reconnaissance Squadron, flying a sensor relay mission, would pick up the signal and relay the signal to a waiting EC-121R, like this one of the 193rd TEWG/Pennsylvania Air National Guard. The EC-121 would call in a strike mission on the area. (R.W. Harrison)

My Cha Chees was an EC-47P of the 361st TEWS. The electronic Skytrains were attached to the 56th SOW from 1972 through 1974. The EC-47s were used to intercept Communist radio traffic throughout the war. (56 TTW/HO)

This EC-47P, on patrol over Cambodia during 1973, was named *Margaret Ann. The EC-47s took part in project Hawk Eye*, intercepting and monitoring VC/NVA radio traffic. (Robert Wilhelm)

An EC-47Q on the ramp at Udorn, Thailand during 1973. Detachments of EC-47s were stationed all over the Southeast Asia area, including South Vietnam, Cambodia, Laos and Thailand. (Robert Wilhelm)

Night flying Forward Air Controllers (FACs), working with the gunships *Nimrod*s or *Zorro*s, always ran the risk of being hit by friendly fire from above. This 23rd TASS O-2A FAC wanted to make sure he was seen and identified. (USAF)

This Cessna O-2A of the 23rd Tactical Air Support Squadron was painted overall Flat Black with Red lettering and warning stripes. The aircraft carries its mission tally on the fuselage side in Red. (Tom Brewer)

This 23rd TASS OV-10A Bronco was equipped with the *Pave Nail* laser illuminator pod under the fuselage. The laser designator in the pod would pinpoint the target for F-4s armed with laser-guided bombs. (56th TTW/HO)

USAF OV-10s were painted in overall Light Gray with White wing upper-surfaces. The aircraft normally carried 2.75 inch rocket pods, containing marker rockets, on the sponson weapons stations and usually had the M60 machine guns removed.

The only jet powered aircraft assigned to the 56th ACW/SOW during the war were the *Scatback* T-39 Sabreliners. The T-39s delivered intelligence and reconnaissance materials, official mail, and VIPs between Clark Air Base, The Philippines and bases throughout Southeast Asia. (56 TTW/HO)

Beginning in 1970, the A-1 *Sandys* were phased out of the SAR escort role, being replaced by A-7D Corsair IIs. The A-7Ds were assigned to the 3rd Tactical Fighter Squadron attached to the 388th TFW at Korat, Thailand. (Lee Bracken via Robert F. Dorr)

A sharkmouthed A-7D Corsair II of the 3rd Tactical Fighter Squadron. The aircraft has the last three digits of the serial number repeated on the nose wheel door in Red. (Hugh R. Muir)

A-7Ds of the 3rd Tactical Fighter Squadron out of Udorn, Thailand, flew support for the rescue units involved in the MAYAGUEZ incident, supporting the CH-53 *Knives* of the 21st SOS. (Lee Bracken via Robert F. Dorr)

Phantoms And Falcons

On 30 June 1975, the 56th Special Operations Wing was transferred on paper to MacDill Air Force Base, Florida, and given a new name: the 56th Tactical Fighter Wing (TFW), new personnel and new equipment. The unit absorbed the assets of the 1st Tactical Fighter Wing (which began conversion to the McDonnell-Douglas F-15 Eagle and moved to Langley Air Force Base, Virginia) and was re-equipped with the best tactical fighter aircraft in the Air Force at the time, the McDonnell-Douglas F-4E Phantom II.

The mission of the 56th TFW was two-fold. First, they were a typical Air Force Tactical Fighter Wing with the standard Tactical Air Command air superiority and weapons delivery mission (the same basic mission that the unit had during 1944/45). As such they had to maintain crews in a combat ready status at all times, ready to deploy if necessary.

Secondly, the Wing was also responsible for the training of air and ground crews on the F-4 Phantom; as such, the 56th became the

The 56th SOW came home from SEA in 1975, becoming a Tactical Fighter Wing based at McdIll AFB, Florida, equipped with the F-4E Phantom II. This F-4E was assigned to the Deputy Wing Commander as indicated by the fin tip with the Phantom silhouettes in the squadron colors. (Ken Buchanen)

The Wing Deputy Commander's F-4E Phantom II on the ramp at McDill Air Force Base. The aircraft carries both the Tactical Air Command badge on the fin and the 56th Tactical Fighter Wing badge on the air intake. (Ken Buchanen)

main Air Force replacement training establishment for F-4E pilots, weapons systems officers (WSOs) and maintenance personnel for all of Tactical Air Command.

As part of this reorganization, the Wing regained its original squadrons plus one new squadron. The 61st, 62nd, and 63rd Tactical Fighter Squadrons (TFS) were all activated at MacDill on the same day as the 56th TFW. They were joined by the 4501st Tactical Fighter Replacement Squadron, which was redesignated as the 13th TFS and attached to the 56th TFW. During October of 1977, the Wing began conversion training from the F-4E to the F-4D. By 12 September 1978, all four squadrons had completed the transition and were declared operational.

During 1979, the 56th TFW began taking delivery of the newest fighter aircraft in the Air Force at the time, the General Dynamics F-16A Fighting Falcon. The Wing would be the first training unit in the Air Force to be equipped with the new fighter and was responsible for training crews for other Tactical Air Command units. Conversion to the F-16A and F-16B aircraft began on 22 October 1979 and was completed on 1 July 1982. While the transition was taking place, the 62nd took part in the 1979 Red Flag exercise at Nellis Air Force Base, Nevada. This was the last time the unit participated in a major exercise flying the Phantom.

The 56th Tactical Fighter Wing had four squadrons during the mid-1970s: the 61st, 62nd and 63rd Tactical Fighter Squadrons, plus the 4501st Tactical Fighter Replacement Squadron (later redesignated the 13th TFS). The Phantoms of the 13th TFS had White vertical fin leading edges and fin tip. (Ken Buchanen)

The 61st TFS flew its last F-4D mission on 19 November 1979, the 62d TFS on 14 November 1980, and the 63rd converted on 1 October 1981. The 13th TFS was transferred to the 432d TFW at Misawa, Japan, on 30 June 1982, being replaced by the 72nd Tactical Fighter Training Squadron (TFTS)

With the conversion of the 56th TFW to the F-16A and B aircraft, the 56th was redesignated as a Tactical Fighter Training Wing on 1 October 1981. With the change in Wing designation, the attached squadrons were also redesignated, becoming Tactical Fighter Training Squadrons. As a TFTS, each squadron serves as an operational training unit. Training includes air-to-air tactics and weapons delivery/gunnery. In addition, student pilots also receive training in air-to-ground tactics at the nearby Avon Park gunnery/bombing range.

Pilot training courses come in several different versions and lengths. One short course is taught to experienced fighter pilots transitioning into the F-16 from other fighter types. A longer (six months) course is taught to younger first tour pilots coming from advanced training units and slated to join a Tactical Air Command F-16 unit.

During 1988, the Wing began receiving the latest model of the Fighting Falcon, the F-16C and F-16D. The 56th TFTW is now training pilots throughout the U.S. Air Force to always be *ALERT AND READY* and to —*BEWARE THE THUNDERBOLT!*

An F-4E Phantom II of the 61st TFS on the ramp at McDill Air Force Base during 1976. The mission of the 56th was air superiority/ground attack and, at the time, the F-4E was considered to be one of the best aircraft in the world for this multi-role mission. (Don Garrett, Jr.)

During October of 1977, the Wing began conversion from the F-4E to the F-4D, while the mission and squadrons remained the same. This F-4D was assigned to COL Charles Cunningham, the Wing Commander and has the fin tip in the squadron colors (from top) Red, White, Blue, Yellow. (Bill Curry via Marty Isham)

An F-4D of the 62nd TFS on the ramp at Nellis Air Force Base, Nevada for the 1979 *Red Flag* exercises. This exercise is the Air Force's most rigorous and realistic training — and the most dangerous. The baggage pod under the port wing carries the badge of the 388th TFW. (Mick Roth)

During October of 1979, the 56th TFW transitioned into what many feel is the best combat aircraft in the world today, the General Dynamics F-16 Fighting Falcon. This formation of 56th F-16As was led by COL Henry Canterbury, the 56th commander. (56th TTW/HO)

F-16B two seaters of the 61st TFTS on the ramp at McDill Air Force Base during 1986. The 56th is responsible for training both air and ground crews on the F-16. The 56th is strictly a training unit, designated the 56th Tactical Training Wing. (56th TTW/HO)

The F-16A Plus was flown by the commander of the 61st AMU during 1988. Although training units, the squadrons of the 56th fly first-line F-16s, fully equipped for standard tactical fighter missions. (Norm Taylor)

Currently, the 56th TTW flies the F-16C and F-16D. This F-16C was flown by COL James Jamerson while he was commander of the 56th TTW. The White square on the intake cover has the badges of all four squadrons surrounding the Wing badge. (56th TTW/HO).

This F-16A was flown by COL Henry Viccellio when he commanded the 56th TFTW in January of 1982. The aircraft was marked with a four color fin cap in squadron colors (from front) Yellow 61st TFTS, Blue 62nd TFTS, Red 63rd TFTS and Green 72nd TFTS. (Norm Taylor)

An F-16A Fighting Falcon of the 62nd Tactical Fighter Training Squadron in flight over Florida. The aircraft carries an AIM-9L Sidewinder air-to-air missile on the wingtip rail. The Falcon carries both the Tactical Air Command badge (on the fin) and the 56th TFTW badge (on the air intake). (Jeff Ethell)

As newer aircraft were brought into the inventory, the Group commander always looked for an aircraft with the numbers "5" and "6" in the serial. COL Ron Fogleman took this F-16A Plus as his personal aircraft during 1983. (Norm Taylor)

Armament technicians load a practice bomblet dispenser under the wing of an F-16B of the 61st TFTS F-16B, using a weapons loading tractor. The first stripe is Yellow with White trim and under it is the Tactical Air Command badge. (Al Lloyd)

An F-16B two seater of 61st Tactical Fighter Training Squadron is hooked up to a starter unit on the ramp at McDill Air Force Base and has both crew boarding ladders in place. The F-16B is a fully combat capable two seat trainer variant of the basic F-16 fighter. (Don Garrett, Jr.)

Aces of the 56th Fighter Group

Name	Squadron	Air/Air	Ground	Total	Name	Squadron	Air/Air	Ground	Total
Francis Gabreski (+six MiGs in Korea)	61	28	2½	30½	Leslie Smith	61 62	6 1		7
Robert S. Johnson	61 62	26 2		28	John Truluck	63	7		7
David Schilling	61 62 63	4 13 6	10½	33½	Mark Moseley	62	6½		6½
					James Carter	61	6		
Fred Christensen, Jr.	62	22		22	Walter Cook	62	6		6
Walker Mahurin (+3 1/2, Pacific)	63	21		21	Lucian Dade	62	6	6	12
Gerald Johnson	61	17		17	George Hall	63	6		6
Hubert 'Hub' Zemke (+2½ w/479 FG)	61 62 63	6 3 7	8½	24½	Cameron Hart	63	6		
					Robert Keen	61	6		6
Joe Powers	61 62	12 2½		14½	Joseph Bennett (+3 with 4FG)	61	5½		5½
Leroy Schreiber	61 62	2 12		14	Frank McCauley	61	5½		5½
					Donovan Smith	61	5½		5½
Felix Williamson	62	13		13	Harold Comstock	63	5		5
Michael Quirk	62	12	5	17	Joseph Egan	63	5		5
Paul Conger	61 63	6½ 5		11½	Steven Gerick	61	5		5
					Norman Gould	62	5		5
James Stewart	61	11½		11½	Joe Icard	62	5		5
Michael Gladych (+16 w/RAF,FAF,PAF)	61	10		10	Evan McMinn	61	5		5
Stanley Morrill	62	10		10	John Vogt (+3 with 356 FG)	63	5		5
Robert Rankin	61	10		10	Eugene O'Neil	62	5		5
George Bostwick	62 63	5 3	6	14	Burton Blodgett	61		5	5
Michael Jackson	62	8	5½	13½	Claude Chinn	63		7	7
Glen Schiltz	63	8		8	Walter Flagg, Jr.	63	2	6	8
Billy Edens	62	7		7	Russell Kyler	61	3	7	10
Frank Klibbe	61	7		7	Randel Murphy	63	2	11	13
Robert Lamb	61	7		7	Vernon Smith	63		5	5